SYNTHETIC MEMBRANES

MMI PRESS SYMPOSIUM SERIES

A series of monographs based on special symposia held at the Michigan Molecular Institute

Editor: Hans-Georg Elias

Volume 1 MOLECULAR MOTION IN POLYMERS BY ESR
Edited by Raymond F. Boyer and Steven E. Keinath

Volume 2 POLYMER COMPATIBILITY AND INCOMPATIBILITY: PRINCIPLES AND PRACTICES
Edited by Karel Šolc

Volume 3 BLOCK COPOLYMERS: SCIENCE AND TECHNOLOGY
Edited by Dale J. Meier

Volume 4 TRANSITION METAL CATALYZED POLYMERIZATIONS: ALKENES AND DIENES
Edited by Roderic P. Quirk

Volume 5 SYNTHETIC MEMBRANES
Edited by Maynard B. Chenoweth

Additional volumes in preparation ISSN 0195-3966

The publisher will accept continuation orders for this series, which may be cancelled at any time and which provide for automatic billing and shipping of each title in the series upon publication. Please write for details.

MMI PRESS SYMPOSIUM SERIES continues the MIDLAND MACROMOLECULAR MONOGRAPHS Series previously published by Gordon and Breach, Science Publishers, Inc. The following titles from that series are available from Gordon and Breach, P.O. Box 786, Cooper Station, New York, NY 10276, U.S.A.:

Volume 1 TRENDS IN MACROMOLECULAR SCIENCE
Edited by Hans-Georg Elias

Volume 2 ORDER IN POLYMER SOLUTIONS
Edited by Karel Šolc

Volume 3 POLYMERIZATION OF ORGANIZED SYSTEMS
Edited by Hans-Georg Elias

Volume 4 MOLECULAR BASIS OF TRANSITIONS AND RELAXATIONS
Edited by Dale J. Meier

Volume 5 POLYMERIC DELIVERY SYSTEMS
Edited by Robert J. Kostelnik

Volume 6 FLOW-INDUCED CRYSTALLIZATION IN POLYMER SYSTEMS
Edited by Robert L. Miller

Volume 7 SILYLATED SURFACES
Edited by Donald E. Leyden and Ward T. Collins

ISSN 0141-0342

Papers presented at the
Sixteenth Michigan Molecular
Institute Meeting held
August 19–22, 1984
in Midland, Michigan

SYNTHETIC MEMBRANES

Edited by
Maynard B. Chenoweth
Michigan Molecular Institute

Published for MMI Press by
harwood academic publishers
chur london paris new york

Published under license by Harwood Academic Publishers

Poststrasse 22
7000 Chur
Switzerland

P.O. Box 197
London WC2E 9PX
England

58, rue Lhomond
75005 Paris
France

P.O. Box 786
Cooper Station
New York, NY 10276
United States of America

Library of Congress Cataloging-in-Publication Data

Michigan Molecular Institute. Meeting (16th : 1984 :
Midland, Mich.)
Synthetic membranes.

(MMI Press symposium series ; v. 5)
Includes index.
1. Membranes—Congresses. I. Chenoweth, Maynard B.
II. Title. III. Series.
TP159.M4M5 1984 660.2′842 86-7604
ISBN 3-7186-0327-6

ISBN 3-7186-0327-6. ISSN 0195-3966. Printed in Great Britain by Bell and Bain Ltd., Glasgow.

Contents

Preface

While membrane technology is potentially beneficial to a great many areas of modern science, its research and applications are vitally important to medicine and the life sciences. It is through the powerful and necessary interrelationship among chemistry, physics and technology that basic human concerns, such as water purification and food processing, may take place.

As Editor of these Proceedings and as a medical scientist, I find myself in the very agreeable position of being able to see examples of both the current state-of-the-art in synthetic membrane technology and research endeavors that, unfortunately, did not prove to be successful. However, I believe the intrinsic value of this collection of papers is its ability to motivate others to initiate and to continue their own investigations.

It is my privilege to extend my heartfelt thanks to a group of people whose efforts made this conference possible. To the Co-Chairmen, Dr. Gordon Fearon and Dr. Charles Reineke, and to the Honorary Chairman, Dr. Hans-Georg Elias, my appreciation for providing the intellectual spirit that helped to make the program such a great success. And, to the Symposium Secretary, Mrs. Sandra Butler, my gratitude for making certain that our temporal needs were more than adequately taken care of.

Maynard B. Chenoweth, M.D.

New Developments in Membrane Processes for Gas Separations

S. ALEXANDER STERN

Development of Chemical Engineering and Materials Science Syracuse University, Syracuse, NY 13210

Membrane separation is now a well-established technology, particularly because of the variety of applications of reverse osmosis and membrane utlrafiltration. Some of these applications are on a substantial scale, such as in the desalination of brackish and sea water. Membrane separation has become economically competitive in these areas because of the development of high-flux asymmetric membranes and the ability to fabricate such membranes in the form of hollow fibers. In recent years, further progress in membrane science has resulted in the successful application of membrane technology to the separation of gases of industrial interest. A number of large-scale membrane plants for the separation of hydrogen from various industrial gas streams and for the recovery of carbon dioxide in extended oil recovery are presently in operation. These membrane processes are competitive with conventional gas separation techniques. The separation of several other important gases by selective membrane permeation, such as of oxygen from air, is being studied in many laboratories. It is anticipated that, in a not too distant future, competitive membrane processes will become available also for the separation of liquid mixtures, such as azeotropes and mixtures or organic compounds. The discussion will outline the engineering and economic factors affecting the development of membrane processes for the separation of both gases and liquids, and will summarize the state-of-the-art in these areas.

I. INTRODUCTION

The development of membrane processes for the separation of gas mixtures has made remarkable progress during the last two decades. The basic concept underlying this technology, namely, the selective permeation of gases through nonporous polymer membranes, appears to have first been proposed by Graham (1) as early as 1866. A full century later, in the early 1960's, membrane processes for gas separations were still being studied largely on a laboratory scale.

Today, over 100 membrane separation plants are in operation in the United States as well as in Europe and in Japan. Most of these plants are for the separation of H_2 from various industrial gas streams, such as in NH_3 and CH_3OH synthesis, in petrolem refining, and in petrochemical operations. However, a number of membrane plants separate CO_2 in enhanced oil recovery (EOR), from natural gas, and from landfill gas, while a few plants separate air to produce 96% N_2 for the "inerting" (blanketing) of fuel tanks. The largest membrane separation plant, which was designed to separate CO_2 from mixtures with hydrocarbons in OER, has been reported to process 50×10^6 cu. ft. ($1.4 \times 10^6 m^3$) of feed gas/day (2). Many other membrane processes for the separation of gas mixtures are being studied at present both on a laboratory and pilot-plant scale, as can be seen from Table I. It has been estimated that the total available market for membrane separation processes could reach $500 million between now and 1990.

The great interest in membrane separation technology is due mainly to the fact that it is potentially energy-efficient. In addition, the required process equipment is simple, compact, and relatively easy to operate and control. Moreover, this equipment is modular and can be easily scaled up or operated by reduced capacity if necessary. Membrane separation became economically competitive in 1961, when Loeb and Sourirajan (3) developed their high-flux, asymmetric membranes of cellulose acetate for water desalination by reverse osmosis. Asymmetric membranes can now be made from many types of polymers, and are being employed for the separation of gases both in sheet form and in the form of composite hollow fibers. However, the greatest impetus to the development of membrane separation technology was given by the energy crisis of 1973, which occured just at the time when this technology had reached a significant degree of maturity.

The objective of the present paper is to summarize some of the charac-

TABLE I

Membrane processes for the separation of gas mixtures

H_2 from ammonia synthesis purge gas
H_2 from methanol synthesis purge gas
H_2 from hydrodesulfurizer purge gas
H_2 from hydrogenator purge gas
H_2 from petrochemical synthesis gas
O_2 and N_2 from air
CO_2 separation in enhanced oil recovery
CH_4 from biomass, landfill, coal mine, and oil field gases
CH_4 from low BTU (CO_2- and N_2-rich) natural gas
He recovery from natural gas
CO_2 and H_2S scrubbing of acid gases
Dehydration of natural gas and of industrial gases
Separation of Kr and Xe from nuclear reactor atmospheres

teristics of membrane processes for the separation of gas mixtures, and to highlight new and promising developments in this field.

II. MECHANISM OF GAS PERMEATION

It is generally accepted that gases permeate through nonporous polymer membranes by a "solution-diffusion" mechanism (4–8). This mechanism can be described with reference to Figure 1, which shows a cross-section through a planar membrane. Application of gas pressure at one interface of such a membrane results in the following sequence of events: (1) solution (absorption) of the gas into the membrane at that interface, (2) molecular diffusion of the gas in and through the membrane, and (3) release of the gas from solution (desorption) at the opposite interface. The term *permeation* is accordingly used to describe the overall mass transport of transport of "penetrant" gas across the membrane, whereas the term *diffusion* refers only to the movement of gas molecules inside the polymer matrix.

In most cases of interest in membrane separations, molecular diffusion is the slowest and, hence the rate-determining step in the permeation process. By contrast, the absorption and desorption steps are so fast that solution equilibrium is usually established at the membrane interfaces. Diffusion of a gas in and through a polymer membrane can usually be described by Fick's two laws. By solving Fick's laws for the desired physical conditions, i.e., for the appropriate initial and boundary conditions, it is possible to formulate the rate of diffusion J as well as the penetrant concentration profile within the membrane. J is also the rate of gas permeation through the membrane since, as mentioned above, diffusion is rate controlling. Crank (8) and Jost (9) have described various methods of solving Fick's equations for different membrane geometries and boundary conditions, for constant and variable diffusion coefficients, and for both transient and steady-state transport.

From a practical viewpoint, it is of interest to determine the rate of gas permeation under *steady-state* conditions, J_s, as well as the dependence of this rate on pressure and temperature. Steady-state is achieved if, at a given temperature, the constant pressures p_h and p_ℓ ($<p_h$) are maintained at the two membrane interfaces, respectively. For example, the following expression can be derived from Fick's first law for the steady-state rate of gas permeation through a planar, isotropic, and homogeneous membrane of thickness δ (4, 5):

$$J_s = \bar{P}(p_h - p_\ell)/\delta, \tag{1}$$

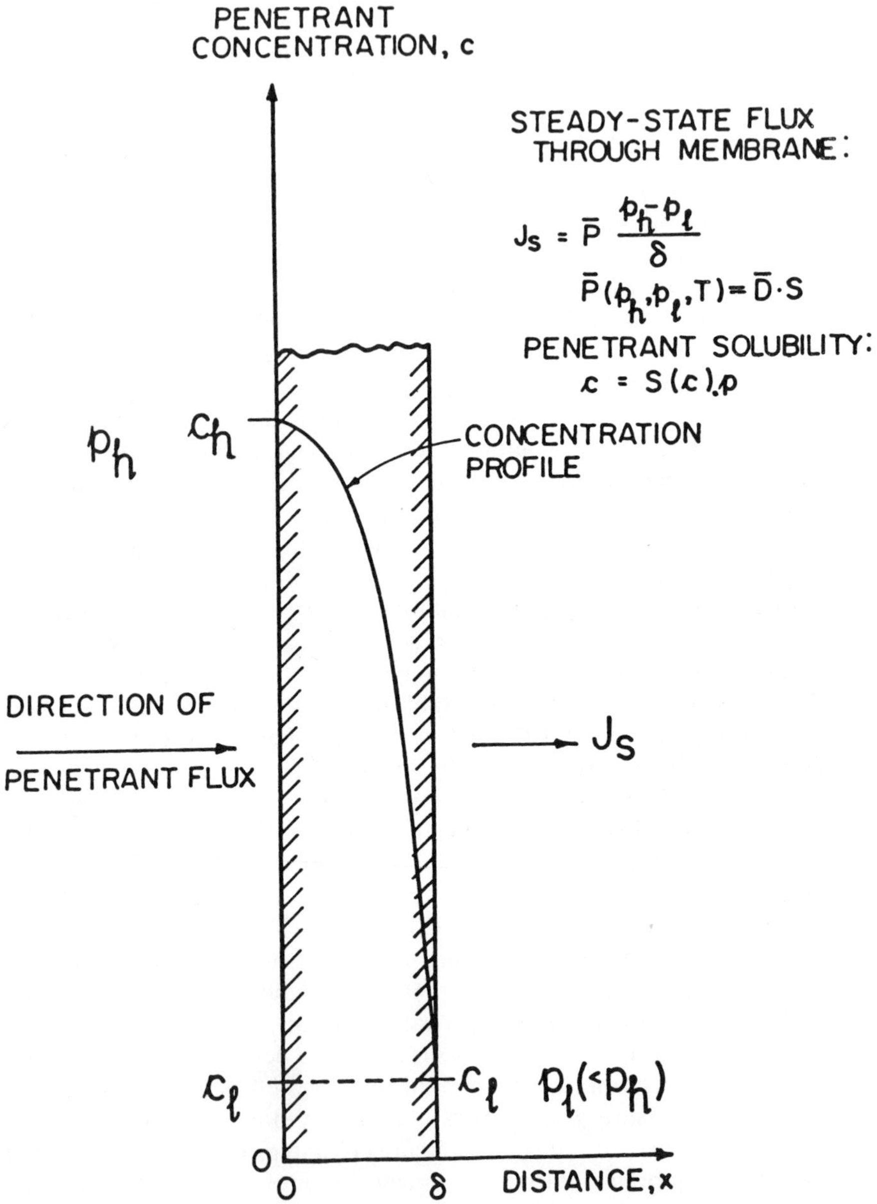

FIGURE 1 "Solution-diffusion" mechanism of gas permation. Cross-section of a non-porous polymer membrane.

where $\bar{P}$ is a mass-transfer coefficient known as the mean "permeability coefficient" or the "permeability"; eqn. (1) refers to a specified temperature and unit area of membrane.

It can be further shown that when $p_h \gg p_\ell$, as is the case in membrane separation processes, the permeability coefficient $\bar{P}$ is a product of a mean diffusion coefficient $\bar{D}$ and a solubility coefficient S_h:

$$\bar{P} = \bar{D} \cdot S_h, \tag{2}$$

where

$$\bar{D} = \frac{\int_{c_\ell}^{c_h} D(c)dc}{(c_h - c_\ell)},$$

$$S_h = c_h/p_h, \tag{4}$$

and where c_h and c_ℓ are the equilibrium concentrations of penetrant dissolved at the two membrane interfaces when the penetrant pressures at these interfaces are p_h and p_ℓ, respectively; and D(c) is the mutual diffusion coefficient for the penetrant/membrane system. In the most general case, $\bar{P}$ depends on the nature of the penetrant and the polymer membrane, on both p_h and p_ℓ, and on the temperature.

Similarly, it can be shown that the steady-state rate of gas permeation through a tubular membrane, such as a capillary or hollow fiber, is given by the relation:

$$J_s = \bar{P}\frac{2\pi L(p_h - p_\ell)}{\ln(R_0/R_I)}, \tag{5}$$

where R_0 and R_I are the effective outer and inner radii of the tube, and L is the length of the tube.

The rate of permeation of a component of a gas mixture is also given by eqns. (1) or (5), but the permeability coefficient $\bar{P}$ may then be a function of composition as well, where p_h and p_ℓ are then the *partial* pressures of that component of the two membrane interfaces.

The *selectivity* of a nonporous membrane toward two different components A and B of a permeating gas mixture is usually expressed in terms of an ideal separation factor, α^*, which is defined by the relation(s):

$$\alpha^*(A/B) \equiv \bar{P}^A/\bar{P}^B \tag{6}$$

If the membrane exhibits a higher selectivity for component A than for B (i.e. at equal partial pressures, A permeates faster through the membrane than B) then $\alpha^*(A/B) > 1$, whereas the opposite is true if $\alpha^*(A/B) < 1$. No separation can occur, of course, when $\alpha^*(A/B) = 1$. The ideal separation factor is, therefore, a separation index similar to the relative volatility used in distillation.

III. PRINCIPLES OF MEMBRANE PROCESS DESIGN

A. General Considerations

Gas mixtures are separated by selective permeation through a polymer membrane in devices known as "permeators". The permeator is the basic membrane separation unit, or stage, and consists essentially of a high-pressure vessel containing a large area of membrane. The vessel is provided with an inlet for the gas mixture to be separated (the "feed") and outlets for two or more product streams. Additionally, the permeator is provided with the necessary piping and valving, and with instrumentation for the control and measurement of flow rates, pressures, and temperature. The compositions of the feed and product streams are also monitored.

The operation of a conventional permeator is illustrated in Figure 2, where the permeator is represented conceptually as a "black box" separated into two compartments by a planar membrane. The feed stream flows through one of the compartments at some suitable pressure p_h, which is usually maintained constant. The opposite compartment is maintained at some lower constant pressure, p_ℓ. A fraction of the feed, known as the "stage cut", is allowed to permeate through the membrane into the low-pressure compartment. The feed stream is thereby separated into two product streams: (a) a permeated (low-pressure) stream enriched in the faster permeating components of the feed, and (b) an unpermeated (high-pressure) stream depleted in these components. The magnitude of the stage cut will depend on the feed flow rate, the membrane area, and the pressure ratio $r = p_h/p_\ell$.

It should be noted that the membrane need not be planar, or in sheet form, as is shown in Figure 2, but can also be, and most often is in the form of a bundle of hollow fibers or capillaries. A diagram of a hollow fiber permeator is shown in Figure 3, where the high- and low-pressure streams are represented as flowing countercurrently to one another. It is seen that a hollow fiber permeator greatly resembles a tubular heat exchanger.

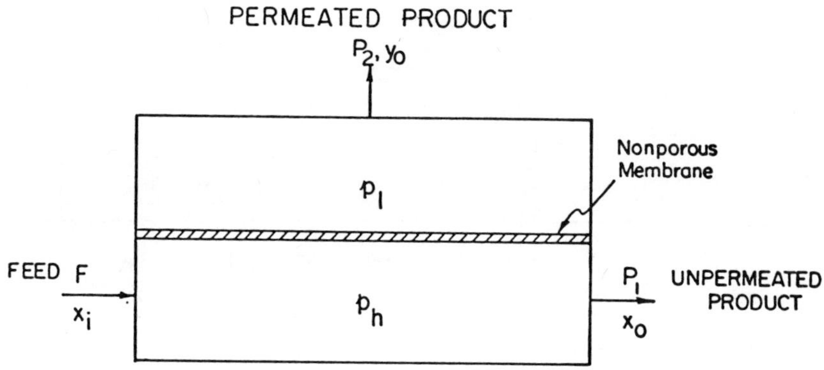

<u>FACTORS AFFECTING PRODUCT COMPOSITIONS AND MEMBRANE AREA</u>

1) Feed Composition
2) Nature of Membrane
3) Temperature
4) Pressures p_h and p_ℓ ($< p_h$)
5) Stage Cut θ ($= P_2/F$)
6) Flow Patterns on High- and Low-Pressure Sides of Membrane

FIGURE 2 Diagram of permeator (single permeation stage).

Given the composition of the feed stream, the process designer must select a membrane with a suitable gas selectivity and permeability. He may be guided in this task by the criteria discussed in a following section. The process designer then wishes to determine the compositions of the product streams and the membrane area requirement for various operating conditions. The extent of separation achievable in a permeator (i.e. in a single permeating stage), and hence the compositions of the product streams, will depend on the following factors (5, 10–12), cf. Figure 2:

Feed composition
Nature of the membrane
Pressures p_h and p_ℓ

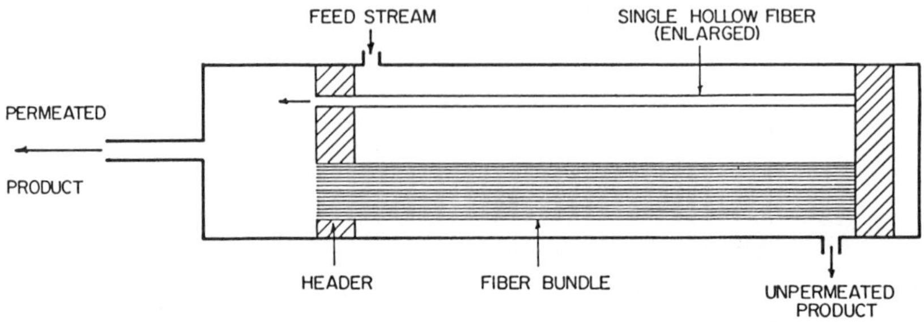

FIGURE 3 Hollow-fiber permeator.

Temperature
Stage cut
Flow patterns of the high- and low-pressure streams in the permeator.

The flow rates of the product streams are determined by the feed flow rate, and by the selected stage cut and pressure ratio $r(=p_h/p_\ell)$. The membrane area also depends on the above factors as well as on the feed flow rate.

The first four factors listed above are expressed by the permeability coefficient $\bar{P}$, whose numerical value must be determined experimentally for the desired conditions or predicted, if possible from suitable correlations. The effects of the last two factors are discussed below.

B. Effect of Stage Cut

Consider the membrane separation of a binary gas mixture of components A and B, where A is assumed to be the more rapidly permeating component. The effect of the stage cut θ on the composition of the permeated product stream (the "permeate") is illustrated in Figure 4 for a hypothetical separation process. The stage cut θ is the fraction of the feed permeating through the membrane; thus, if F, P_1, and P_2 are the total flow rates of the feed, unpermeated, and permeated streams, respectively, then $\theta \equiv P_2/F$.

It is seen from Figure 4 that, for specified values of $\alpha^*[=\bar{p}(A)/\bar{P}(B)]$ and of $r(\equiv p_h/p_\ell)$, the highest concentration of component A in the permeate is obtained in the limit $\theta \rightarrow 0$, i.e. when an infinitesimal fraction of the feed permeates through the membrane. As θ is increased, the concentration of A in the permeate decreases. When $\theta = 1$, the other

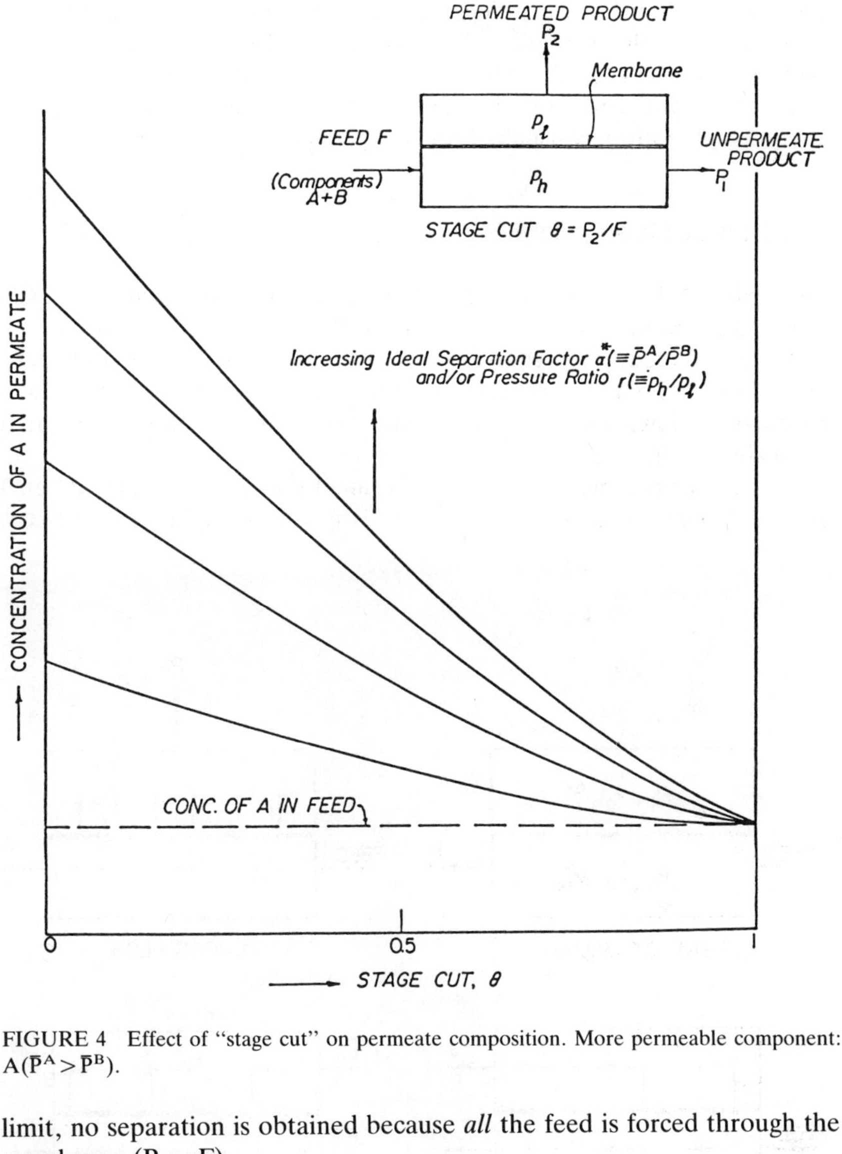

FIGURE 4 Effect of "stage cut" on permeate composition. More permeable component: $A(\bar{P}^A > \bar{P}^B)$.

limit, no separation is obtained because *all* the feed is forced through the membrane ($P_2 = F$).

The process designer is thus faced with an optimization problem. In order to increase the concentration of component A in the permeate (at a given α^* and r) he must decrease θ, thereby decreasing also the permeate flow rate, and vice versa. At a given θ, the concentration of A in the permeate can be increased by increasing α^* or r, or both these parameters. The value of α^* depends on the chemical composition and morphology of the membrane. The possibility of increasing α^* for a given pair of penetrant gases depends, therefore, on the ability of polymer chemists to synthesize more highly gas-selective membranes. The value of the pressure ratio r is usually limited by economic considerations.

C. Effect of Flow Pattern

The extent of separation achievable in a single permeator and the required membrane area depend also on the flow patterns of permeated and unpermeated gas streams, relative to one another, in the permeator (5, 10–12). Four such patterns, for "perfect mixing", cross flow, concurrent flow, and countercurrent flow, are illustrated diagrammatically in Figure 5.

In the "perfect mixing" case it is assumed that the unpermeated high-pressure) gas stream in the permeator is mixed so rapidly that its com-

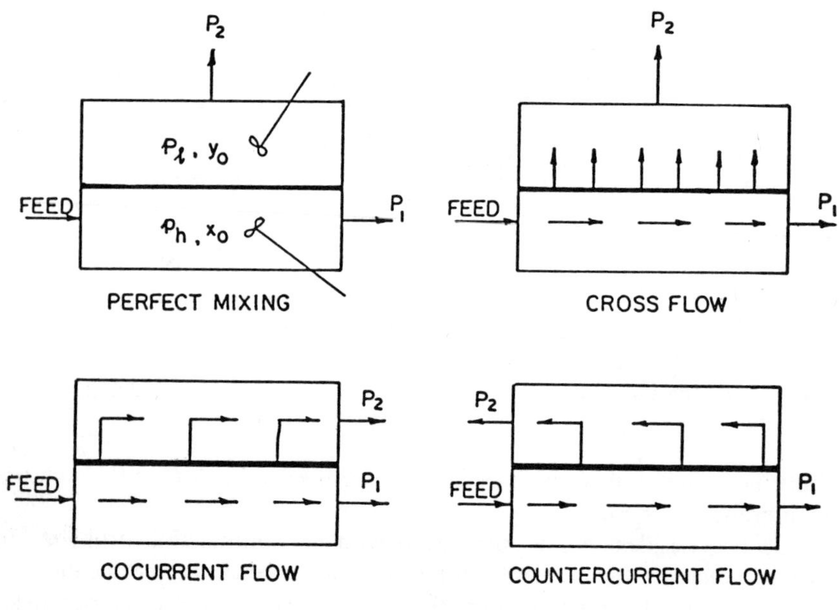

FIGURE 5 Flow patterns in single permeation stage. P_1 = Unpermeated Product; P_2 = Permeated Product.

position at *all points* in the permeator is the same as that at the permeator outlet. The same assumption is made for the permeated (low-pressure) stream. A step-wise change in the feed composition occurs here at the permeator inlet, as is the case in the ideal continuous stirred tank reactor (CSTR). The other flow patterns illustrated in Figure 5 are self-explanatory. In these cases, the composition of both the permeated and unpermeated streams vary continuously from the inlet to the outlet of the permeator, both streams becoming depleted in the more rapidly permeating component(s).

The effect of the flow pattern in a permeator on the separation of a hypothetical mixture of components A and B is shown as a function of stage cut in Figure 6. The plot is for given values of α^* and of the pressure r, and A is again assumed to be the more rapidly permeating component. As is indicated in Figure 6, countercurrent flow is the most efficient flow pattern at any given stage cut $0<\theta<1$, in that it yields the highest concentration of A in the permeate. It can also be shown that countercurrent flow requires the smallest membrane area. By contrast, "perfect mixing" is the least efficient flow pattern in these respects. Other flow patterns yield intermediate results.

The difference in the effects of the above flow patterns on the extent of separation and membrane area requirements is more pronounced the larger the values of α^* and r. However, all flow patterns yield the same results in the stage-cut limits $\theta \rightarrow 0$ and $\theta \rightarrow 1$.

The factors affecting the separation of gas mixtures by selective permeation are now well understood. The performance of permeators can be predicted with confidence for any desired operating conditions from suitable mathematical models (5, 10, 13–24), if the pertinent permeabilty coefficients are known.

IV. NEW CONCEPTS IN MEMBRANE PROCESS DESIGN

A. Motivation

At the present time, the number of membrane processes for gas separations which are economically viable is very limited because of the lack of membranes with a sufficiently high selectivity and permeability toward a spectrum of different gases. This situation is not surprising since most available membranes of commercial origin were developed for purposes other than for gas separation, e.g. for packaging. Considerable work aimed at the synthesis of more highly gas-selective membranes for a variety of separation processes is currently in progress in many labor-

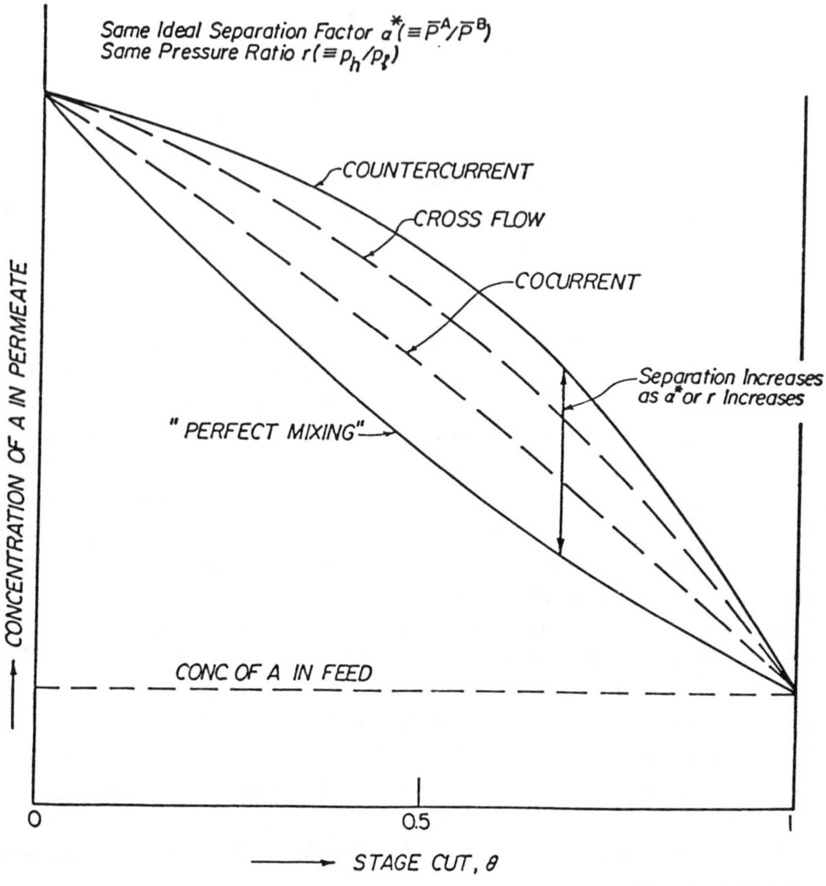

FIGURE 6 Effect of "stage cut" on permeate composition for different flow patterns. More permeable component: $A(\bar{P}^A > \bar{P}^B)$.

atories. However, the extent of separation that can be achieved in a membrane process can also be considerably enhanced in some cases by several process design techniques, using already available membranes.

A well-established technique of increasing the extent of separation in a membrane process is to connect two or more permeators *in series* to form a countercurrent cascade, cf. Figure 7. The permeated stream from each permeator is recompressed and used as feed for the succeeding permeator,

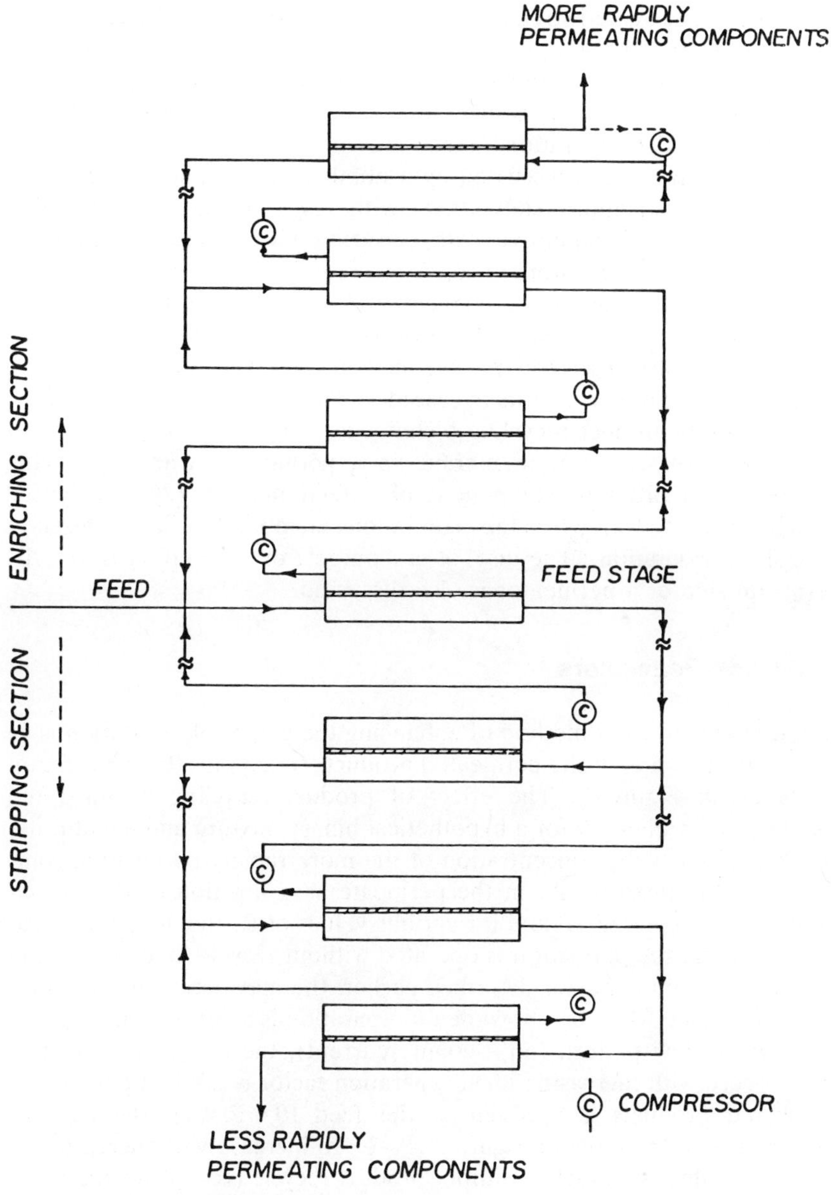

FIGURE 7 Countercurrent permeation cascade.

while the unpermeated stream is returned and mixed with the feed to the preceding permeator. Large cascades using porous barriers as separation elements are presently used in several countries for the separation of uranium isotopes by gaseous diffusion. The principles of cascade design, as applied to the separation of gas mixtures with nonporous polymer membranes, have been discussed by a number of investigators (5, 25).

Economic evaluations show that, with few exceptions, permeation cascades are not competitive with conventional separation techniques even if only two permeators in series are required. Of course, a number of permeators can always be connected *in parallel* in order to increase product output. Two relatively new concepts in membrane process design, which are discussed below, may prove to be more economical than the staging of permeators in cascades. These concepts involve (1) the use of various product recycling techniques, and (2) the use of two (or more) different types of membrane in a permeator, each type being selective to a different component of a feed mixture (26). A special recycle permeator known as the "continuous membrane column" deserves special consideration. The use of a "purge" or "sweep" gas on the permeate side of a permeator is also discussed.

B. Recycle Permeators

A potentially efficient method of enhancing the extent of separation is to recycle a fraction R of the permeated product stream into the feed stream as shown in Figure 8. The effect of product recycling is illustrated qualitatively in Figure 9 for a hypothetical binary mixture and membrane, The figure shows the concentration of the more rapidly permeating component of the mixture, A, in the permeate as a function of the *overall* stage cut $\theta = (1 - R)P_2/F$ and for various values of R. For $R = 0$, i.e., for the case where the permeator is operated without recycle, one obtains the usual permeation behavior described earlier: the maximum concentration of A is obtained when $\theta \rightarrow 0$, while no separation is achieved when $\theta \rightarrow 1$. For a given flow pattern (e.g., countercurrent), the concentration of A will increase with increasing ideal separation factor α^*. When part of the permeated products is recycled to the feed $(0 < R < 1)$, the concentration of A in the plots of Figure 9 is seen to increase with increasing R above the values obtainable without product recycle $(R = 0)$ for the same α^* and r. For large recycle fractions, R, the plot of concentration of component A in the permeated product tends to develop a sharp maximum as R is increased.

The effectiveness of recycling is illustrated quantitatively in Figure 10

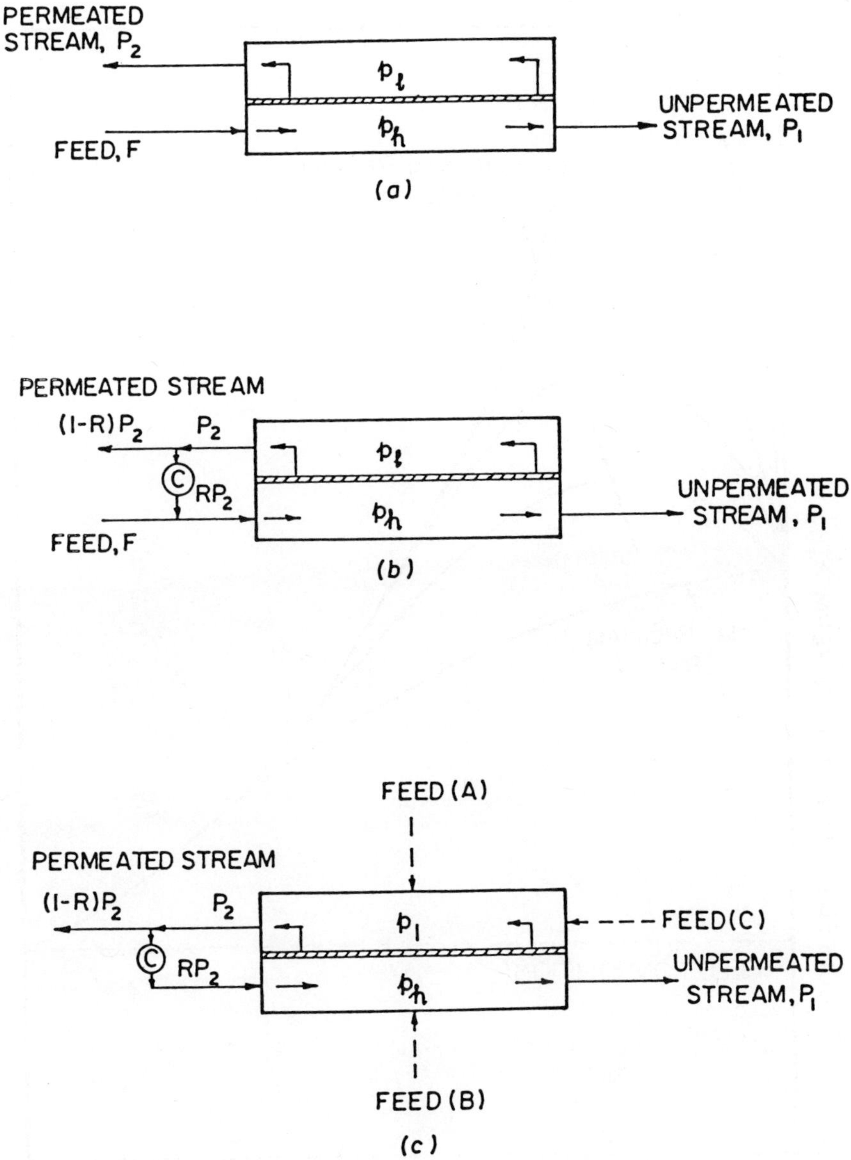

FIGURE 8 Recycle permeators (a) Permeator without permeate recycling; (b) Recycle permeator; (c) Feed inlet points in recycle permeator.

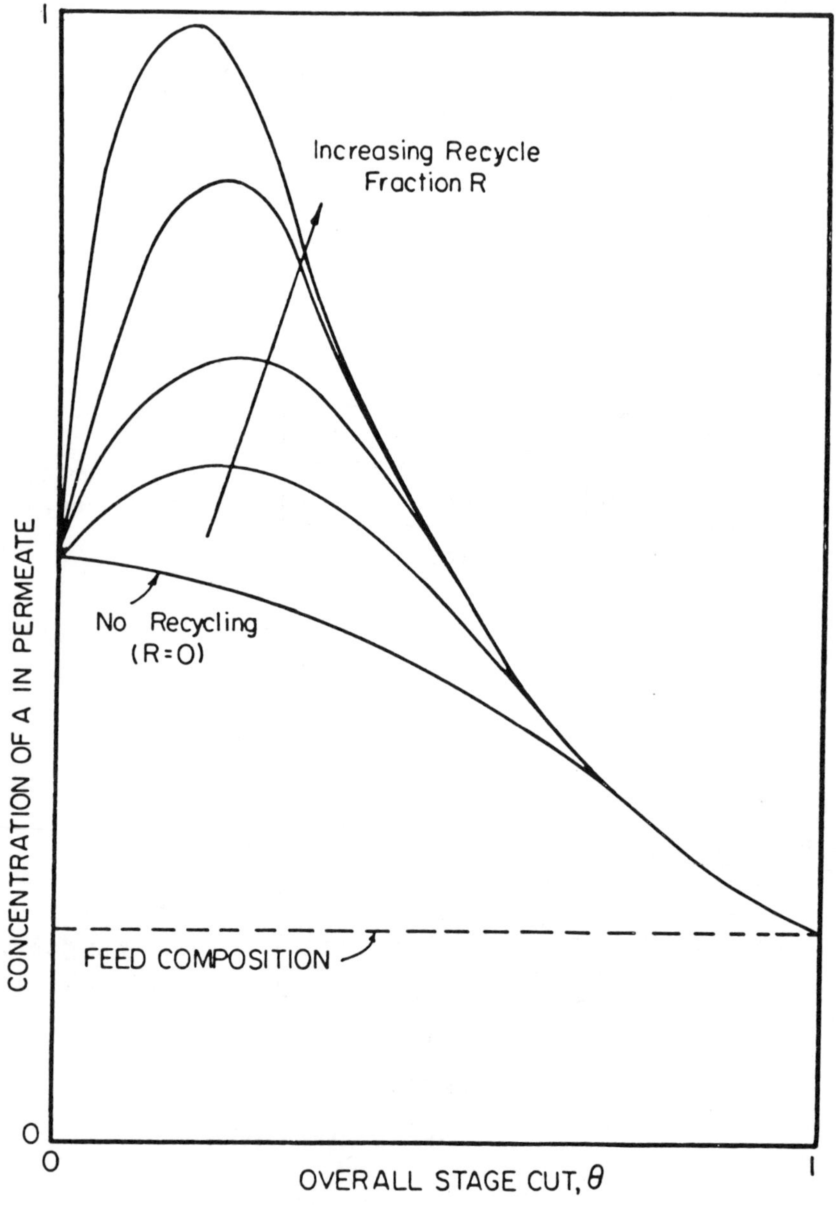

FIGURE 9 Effect of "stage cut" and permeate recycle fraction on permeate composition. More permeable component: $A(\bar{P}^A > \bar{P}^B)$.

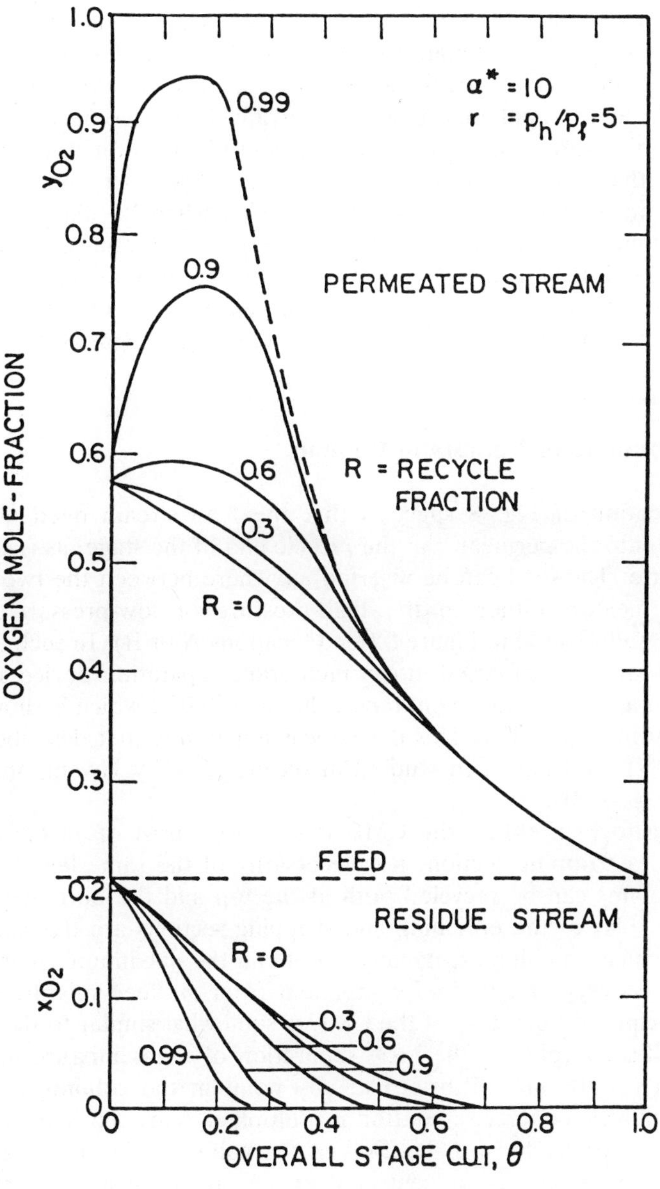

FIGURE 10 Air separation in a recycle permeator. Effect of overall "stage cut" on permeate composition. Countercurrent Flow.

for the case of air separation. In this example, air is taken to be a binary mixture of 20.9 mol % O_2 and 79.1 mol % N_2. It is also assumed that a membrane with an ideal separation factor, $\alpha^*(O_2/N_2)$, of 10 is used and that the pressure ratio r is 5; the assumed value of α^* cannot be obtained with any of the solid polymer membranes available at present, but should be achievable with some "facilitated transport" membrane. Figure 10 shows that the O_2 concentration in the permeated product stream passes through a maximum at a stage cut $\theta = 0.1 - 0.3$ when $R \lesssim 0.6$, and that this maximum increases very rapidly with increasing R values. However, the power and capital investment costs of the permeation process also increase very rapidly with increasing R, because of the increase in the volume of gas recycled and in membrane area requirements (see Figure 11).

C. The Continuous Membrane Column

An examination of Fig. 8 suggests that the feed stream need not be introduced into the permeator at the recycle end of the stage, as is shown in that figure. The feed can be inserted anywhere between the two ends of the permeator, either in the high-pressure or low-pressure compartment, as illustrated in Figure 8 (feed locations A or B). In such cases, the permeator is transformed into a membrane separation device sometimes called a "continuous membrane column" (CMC), which is shown in more detail in Figure 12a. This device was apparently first described by Pfefferle (27), and has been studied in recent years by Hwang and coworkers (e.g., (28)).

Referring to Figure 12a, the CMC is seen to consist of an enriching section and a stripping section, not necessarily of the same length. The product streams can be recycled both at the top and the bottom of the CMC. The roles of the enriching and stripping sections are the same as those in distillation columns, namely, to obtain the maximum separation as well as recovery of a desired component from the feed mixture.

The principle of operation of the CMC is somewhat similar to that of a thermal diffusion column: the local separation of a gas mixture due to permeation (or thermal diffusion) at any point in the column may be small, but this elementary separation is multiplied many times as the gas mixture flows axially toward the ends of the column. The more rapidly permeating component is concentrated on the low-pressure side of the CMC in the low-pressure gas stream.

The extent to which a gas mixture can be separated in a CMC will depend on the recycle fraction R for a given membrane (i.e. for a given

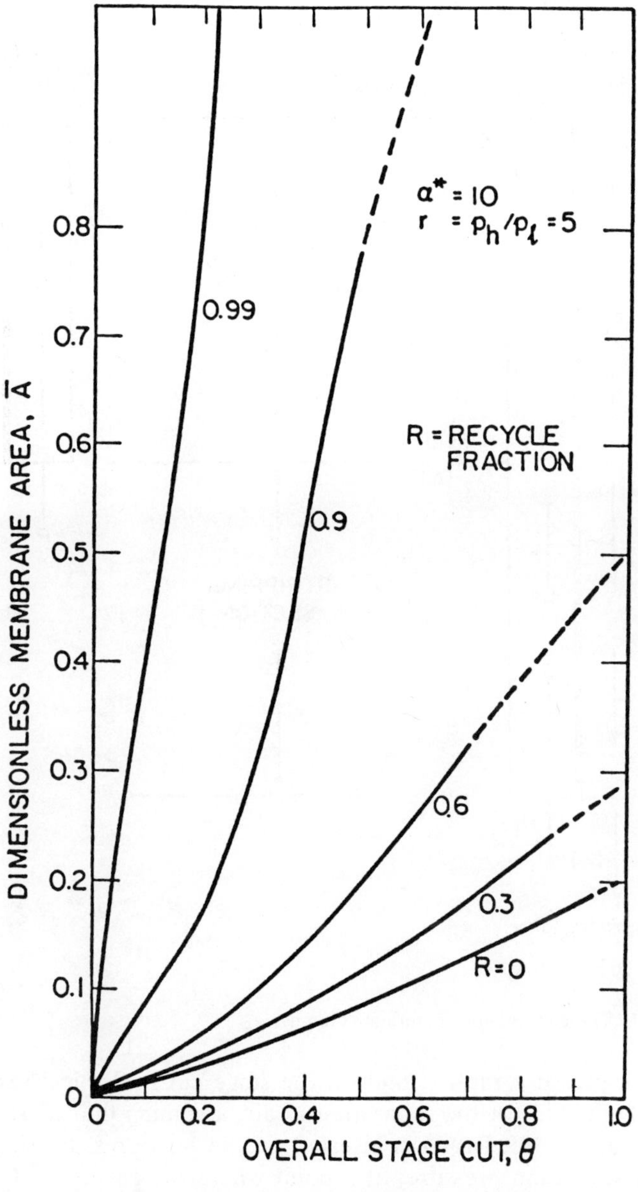

FIGURE 11 Air separation in a recycle permeator. Effect of overall "stage cut" on membrane area requirements. Countercurrent Flow.

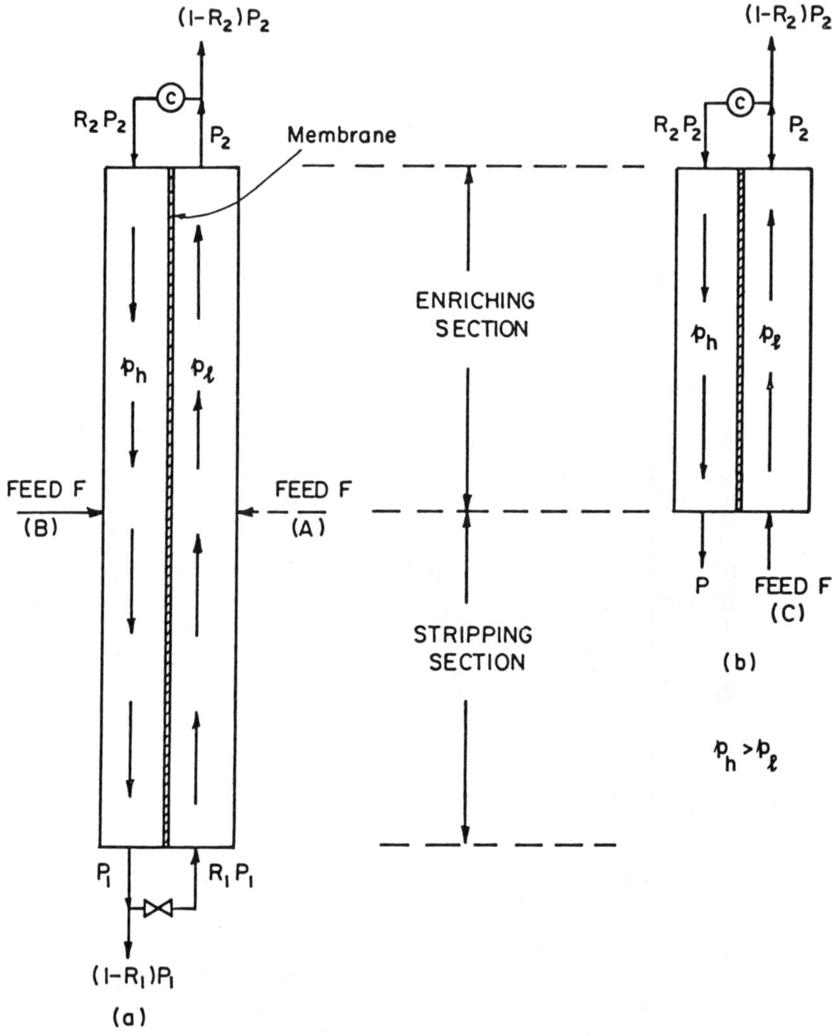

FIGURE 12 The continuous membrane column.

α^*), on the pressure ratio r, and on the stage cut θ. Typical results are shown in Figure 13 for the separation of air, assuming that α^* (O_2/N_2) = 10 and r = 5. The feed (air) is inserted on the high-pressure side of the CMC and at column pressure, at a point where the gas in the CMC has the same composition as the feed.

It is seen from Figure 13 that the concentration of O_2 in the permeated product stream does not exhibit a maximum as was the case in the recycle

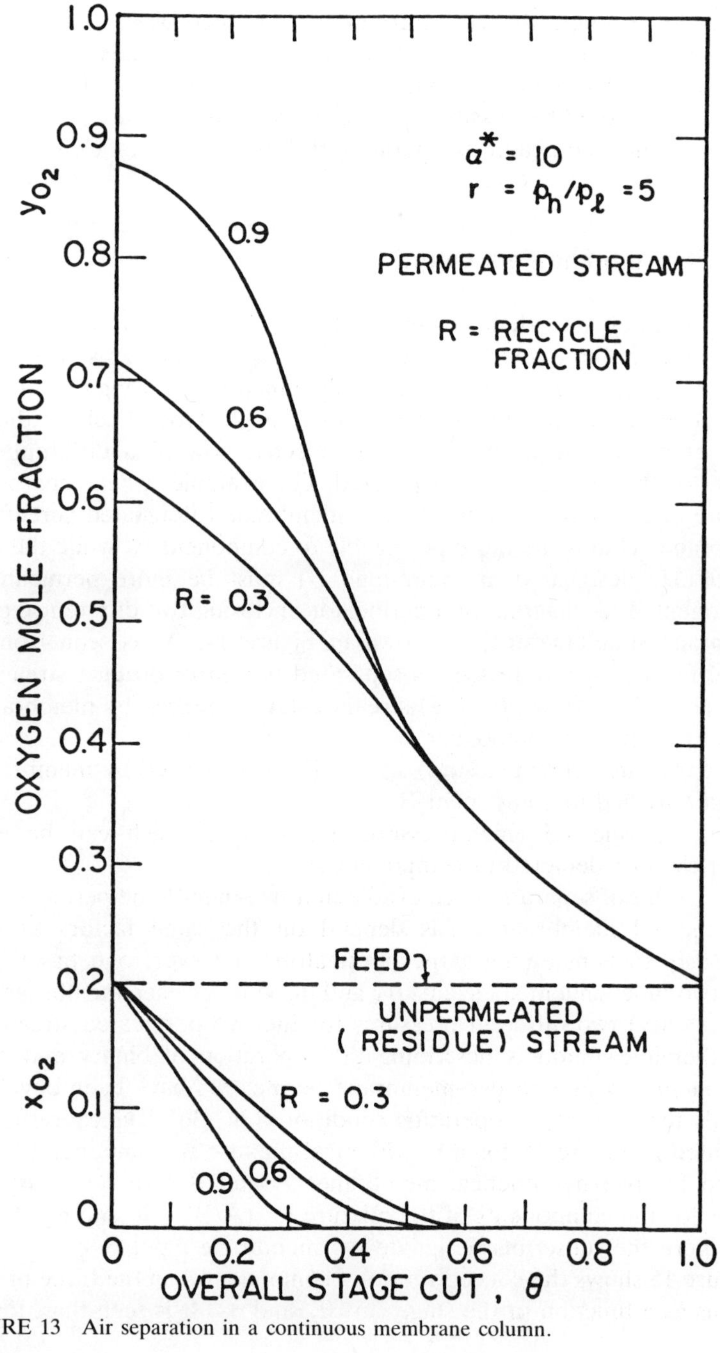

FIGURE 13 Air separation in a continuous membrane column.

permeator. Rather, the concentration of O_2 increases continuously *at constant R* as θ is decreased. The concentration of O_2 also increases *at constant* θ as R is increased, but only at lower values of θ. Therefore, in principle, it should be possible to separate a binary mixture in a CMC to almost any desired extent by operating this device at a sufficiently high R and low θ.

D. Multimembrane Permeators

Onno, Kimura, and their coworkers (29–33) have made the ingenious suggestion that the extent of separation of a binary gas mixture by selective permeation can be substantially enhanced by using two different types of polymer membrane instead of a single type. Each of the two types of membrane is chosen to be selective toward a different component of the mixture to be separated. For example, when separating a mixture of components A and B, one membrane (designated hereafter as membrane A) must be more permeable to component A, while the other membrane (designated as membrane B) must be more permeable to component B. A diagram of a permeator enclosing two different types of membrane simultaneously is shown in Figure 14. As is seen from the figure, the high-pressure feed is separated into *three* product streams:

a) A permeated (low-pressure) stream $P_2(A)$ produced by membrane A and enriched in component A;
b) A permeated (low-pressure) stream $P_2(B)$ produced by membrane B and enriched in component B;
c) An unpermeated (high-pressure) stream P_1, which can be either enriched or depleted in component A.

The extent of separation achievable in a two-membrane permeator and the required membrane areas depend on the same factors as listed previously for single-membrane permeators; however, one has to consider two independent stage cuts θ_A and θ_B, one for each membrane, and (if necessary) two different pressures for the two permeated streams.

Mathematical models describing the separation of binary and multicomponent mixtures in two-membrane permeators have been developed recently for a variety of operating conditions (34–36). Typical results are presented in Figure 15 for a hypothetical mixture of components A and B, and for two hypothetical membranes A and B with reversed selectivities for the components of the mixture [$\alpha_A^*(A/B) = 10$ and $\alpha_B^*(A/B) = 0.1$, where the subscripts designate the membrane type].

Figure 15 shows the concentration of component A in the three product streams as a function of the stage cuts θ_A and θ_B. It is seen that, for any

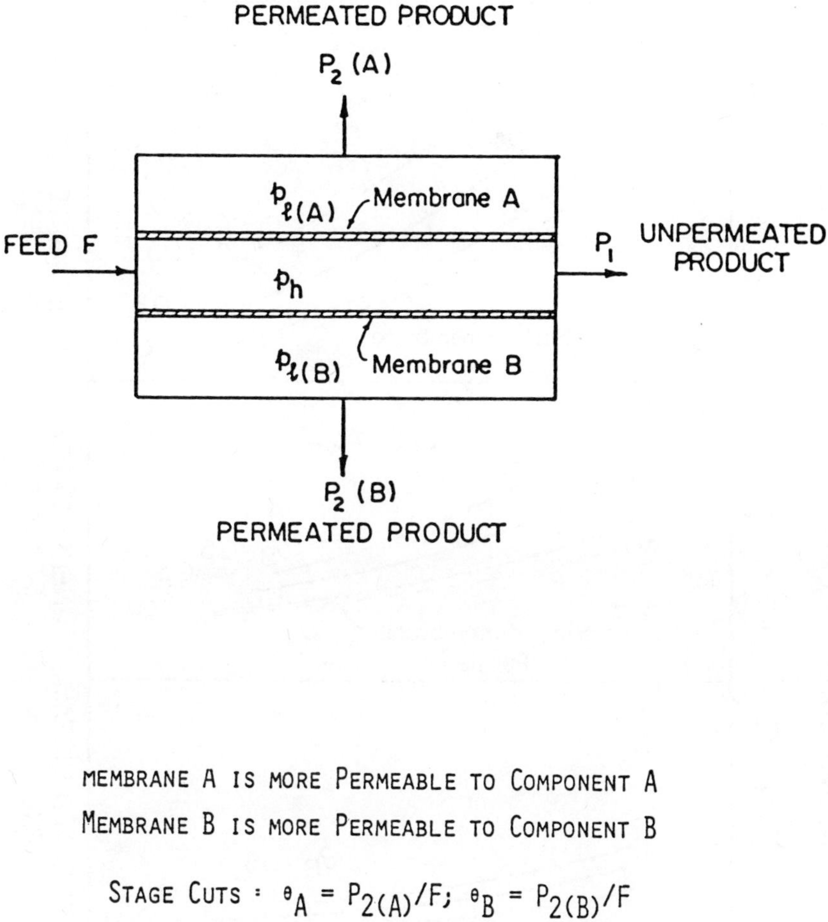

FIGURE 14 Diagram of permeator with two different types of membrane.

given value of θ_A, the concentration of component A increases as θ_B is increased, particularly in the permeate from membrane A. This behavior is due to the fact that, as θ_B increases, more component B permeates through membrane B. Therefore, the unpermeated (high-pressure) stream is enriched in component A; the permeate from membrane A is then also enriched in that component. Figure 15 also shows that a much higher separation can be achieved in a two-membrane permeator than in

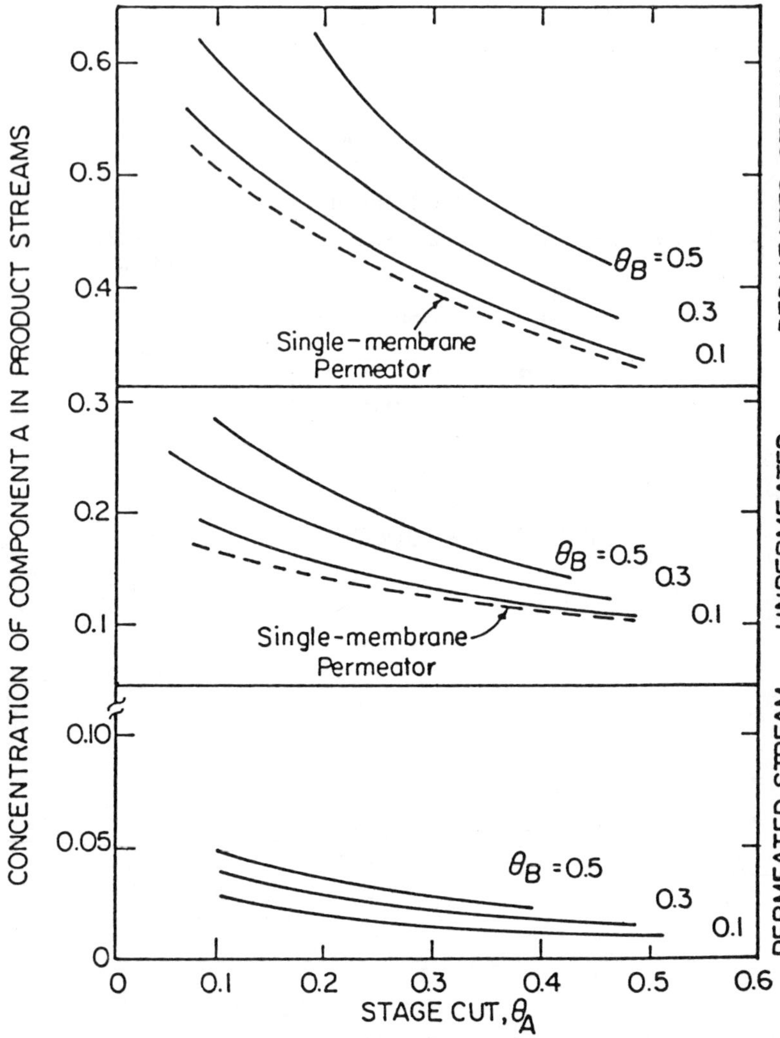

FIGURE 15 Effect of "stage cuts" on compositions of product streams from a two-membrane permeator. $\alpha_A^* = 10$; $\alpha_B^* = 0.1$; $r_A(\equiv p_h/p_{l(A)}) = 5$; $r_B(\equiv p_h/p_{l(B)}) = 5$ "Perfect mixing" conditions. Concentration of component A in feed: 21 mole-%.

a single-membrane permeator provided with membrane A, provided that $\sigma^*_B(A/B)$, θ_B, and r_B are sufficiently large (r_B is the ratio of pressures $p_h/p_{\ell(B)}$ on the two sides of membrane B).

Two-membrane and single-membrane permeators also differ in other respects. For example, the most efficient mode of operating a two-membrane permeator, i.e. the mode yielding the highest degree of gas separation and requiring the smallest membrane area, could be either under countercurrent flow or "perfect mixing" conditions, depending on the relative values of the stage cuts θ_A and θ_B (see Figure 16). By contrast, countercurrent flow is always the most efficient mode of operating a single-membrane permeator.

E. Permeators with Purge Streams

A concept which is not new, but which seems to have received relatively little attention, is the use of a "purge" or "sweep" gas on the permeate (low-pressure) side of a permeator. This procedure would decrease the partial pressures of the permeate components, and thus increase the rates of permeation of these components through the membrane. A theoretical analysis (23) has shown that, with the proper amount of purging, the membrane area requirement would be significantly reduced with only insignificant dilution of the permeated product. If the permeator is operated countercurrently, a small fraction of the feed stream could be used as purge gas. The best results are obtained by purging with an easily condensable gas: not only is the membrane area then reduced, but the dilution of the permeated product can be prevented by condensing out the purge gas (23).

V. MEMBRANE SELECTION

A. Synthetic Polymer Membranes

The "heart" of any permeation process is the membrane that will perform the desired separation. If the permeation process is to compete with conventional separation techniques from an economic standpoint, the membrane must exhibit the following properties:

1. A high permeability towards a specific component (or components) to be separated from a gas mixture.
2. A high selectivity for this component, that is, a high permeability relative to the other components of the mixture.
3. Chemical inertness and physical stability.

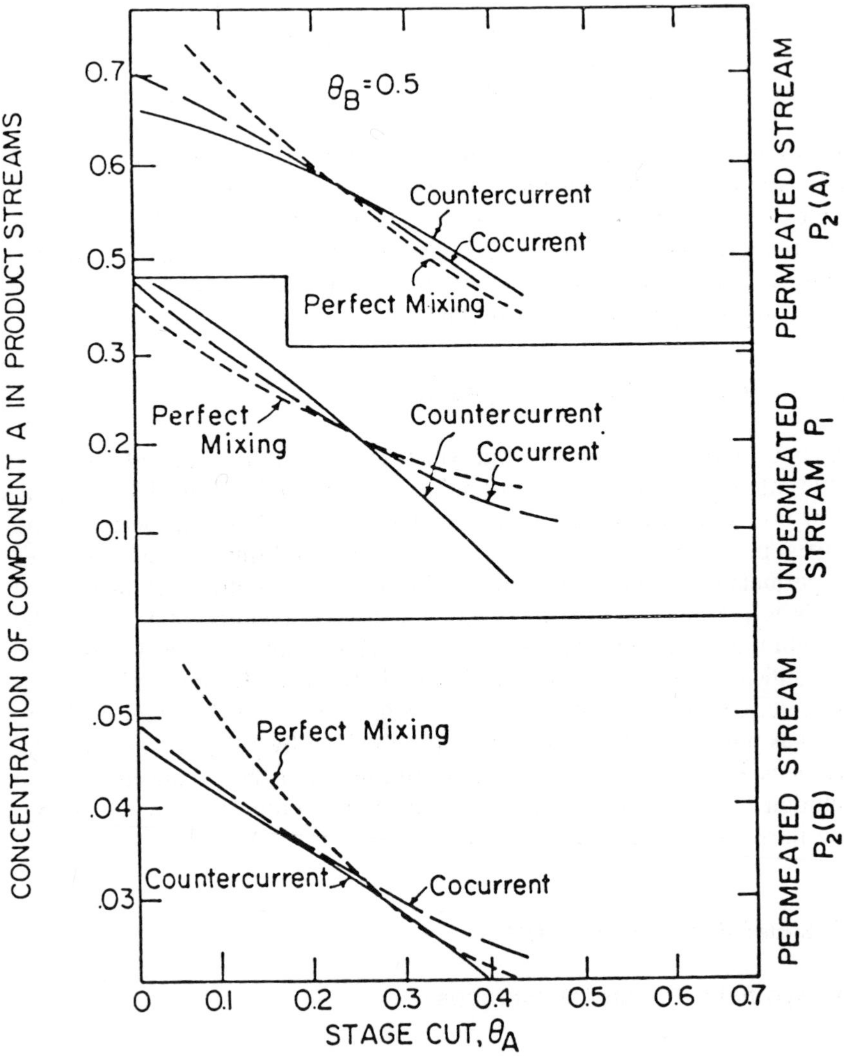

FIGURE 16 Effect of flow pattern on compositions of product streams from a two-membrane permeator. $\alpha_A^* = 10$; $\alpha_B^* = 0.1$; $r_A(\equiv p_h/p_{l(A)}) = 5$; $r_B(\equiv p_h/p_{l(B)}) = 5$ "Perfect mixing" conditions. Concentration of component A in feed: 21 mole-%.

4. Continuity, that is, absence of pinholes or other mechanical defects.

A high permeability and selectivity are required in order to minimize capital investment and power costs (the selectivity must be sufficiently high to yield, at least, the desired product compositions at a practical value of the pressure ratio r). The necessity for chemical inertness and physical stability is self-explanatory. Membrane continuity is required in order to obtain the maximum selectivity characteristic of the membrane. The presence of pores or pinholes in the membrane will greatly reduce, and even destroy, the selectivity for the desired components, because of Knudsen flow or other types of gas transport through such imperfections.

Unfortunately, nature is unkind to the designer of membrane separation processes because available polymer membranes that are highly gas-selective exhibit a low *intrinsic* permeability, and vice versa. This is illustrated in Table II, which lists permeability coefficients and ideal separation factors for He, N_2, and CH_4 in a number of membranes. An examination of this table shows that the membranes which are highly permeable are most often in the rubbery polymer state at the indicated temperature, whereas the membranes which are highly selective are in the glassy state.

The problem of producing a gas separation membrane that exhibits *both* a high permeability and a high gas-selectivity was solved in the early

TABLE II

Permeability coefficients and separation factors for He, N_2, and CH_4 in different polymers

Polymer	Temperature, T(°C)	Permeability coefficient, $\bar{p}(He) \times 10^{10*}$	Ideal separation factor, $\alpha^*(A/B)$ He/N_2	He/CH_4
Poly(1-trimethylsilyl-1-propyne)	25	4100	1.95	0.98
Silicone rubber	30	230	1.5	0.39
Nitrile silicone rubber	30	79	3.8	0.79
Ethyl cellulose (ethocel 610)	25	35.6	10.8	4.8
Polycarbonate	30	67	15	19
Teflon FEP	30	62	25	44
Polystyrene	30	35	16	15
Viton A	30	17	55	110
Polyamide (Nylon 6)	25	2.93	98.8	—
Poly(ethylene terephthalate) (Mylar)	25	1.00	167	167
Polyvinyl fluoride	25	0.97	231	171
Poly(vinylidene chloride) (Aran)	25	0.066	366	263

* $\bar{p}(He)$ is in units of cm^3 (STP)·cm/(s·cm^2·cm Hg)

1960's by Loeb and Sourirajan (3). These investigators developed the now well-known "phase inversion" method for the formation of high-flux, asymmetric membranes (12, 37–41). While orginally intended for the desalting of saline water by the reverse osmosis process, asymmetric membranes are now used also for gas separations.

A cross-section through an asymmetric membrane is shown in Figure 17(a). The membrane consists of a microporous layer (100–200 μm-thick), one side of which is covered with a very thin (0.1–1.0 μm), *nonporous* "skin" or surface layer. The separation of gas mixtures through an asymmetric membrane occurs in its skin, while the microporous substrate provides the membrane with mechanical strength. The rate of gas permeation through such a membrane will be very high because it is inversely proportional to the *effective* membrane thickness, which is that of the nonporous skin. The gas-selectivity of an asymmetric membrane can also be high if, as mentioned above, the membrane is in the glassy state. Asymmetric membranes were first prepared from cellulose acetate, but can now be made from a variety of polymeric materials.

It should be noted that asymmetric membranes may contain solvent and water from the casting and coagulation stages in their preparation, and therefore must be carefully dried before being used as gas separation barriers. Otherwise, the slow removal of solvent and water from the membranes by the permeating gases will cause a continuous increase in the thickness of their dense skin, due to the gradual collapse of the microporous substrate. This will result, in turn, in a corresponding decrease in the membrane permeability. Several methods are now available for drying asymmetric membranes without increasing their skin thickness (42–44).

Another method of preparing an asymmetric membrane is to *laminate* an ultrathin (0.05–0.1 μm), nonporous film to a much thicker (100–200 μm) microporous backing, or support layer, as is shown in Figure 17b. The ultrathin film and backing of this composite membrane, which is sometimes called a "thin-film laminate", have the same functions as the skin and microporous substrate, respectively, of an asymmetric membrane of the type described above. However, the ultrathin film and its backing are made of different materials, whereas the skin of an asymmetric membrane is an integral part of its substrate. As a result, the properties and performance of a thin-film laminate can be better controlled, in principle, than those of a "skinned" membrane. Unfortunately, the preparation of ultrathin films is a delicate operation which cannot be readily automated or scaled up (12, 40–41, 45–49).

A third method of preparing an asymmetric membrane is by *coating* a thin nonporous film directly on the surface of a suitable microporous backing. This coated-film composite differs from a thin-film laminate in that the coating penetrates into the pores of its backing material, as is illustrated in Figure 17c. Consequently, the coating is strongly bonded to its backing. A variety of techniques of preparing coated-film composites have been described in the literature, cf. refs. (12) and (39–41). An important advantage of coated-film composites is that these can be prepared so that the separation of gas mixtures permeating through such membranes occurs either in the coated film or in its backing (e.g. if the backing is itself a "skinned" membrane).

Some asymmetric membranes, such as the "skinned" membranes and the coated-film composites, can be produced in the form of *hollow fibers*

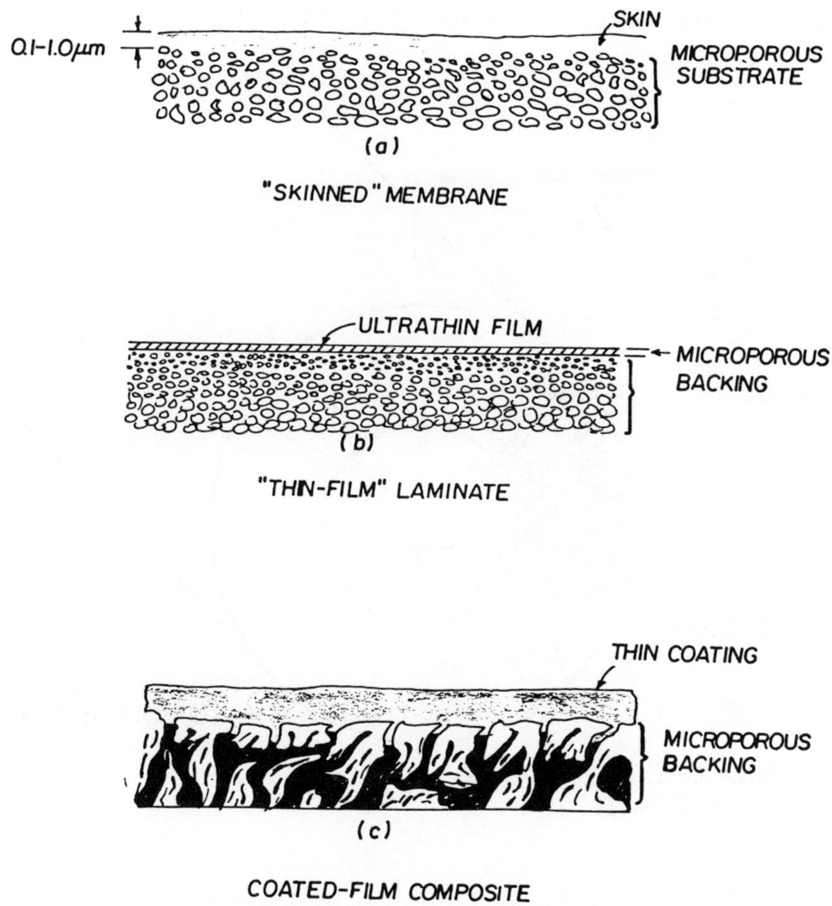

FIGURE 17 Asymmetric membranes.

with inner diameters varying from a few tenths to a few hundreds μm. A cross-section of a "skinned" asymmetric hollow fiber is presented in Figure 18; the nonporous skin is shown on the outer surface of the hollow fiber, but could as well be formed on its inner surface. Hollow-fiber permeators (see Figure 3) are the most efficient membrane separation devices because they can pack much larger membrane areas per unit volume ($>10{,}000\,m^2/m^3$) than permeators provided with membranes in sheet or tubular form ($\lesssim 1{,}000\,m^2/m^3$). Hollow-fiber permeators do not require membrane supports and separators, as do permeators utilizing sheet membranes (plate-and-frame and spiral wound permeators). Moreover, hollow fibers can withstand pressure differences as high as 100 atm. across their walls if they are sufficiently rigid, e.g. if in the glassy state, and if the wall thickness is adequate [O.D./I.D. ≈ 2, ref. (3)].

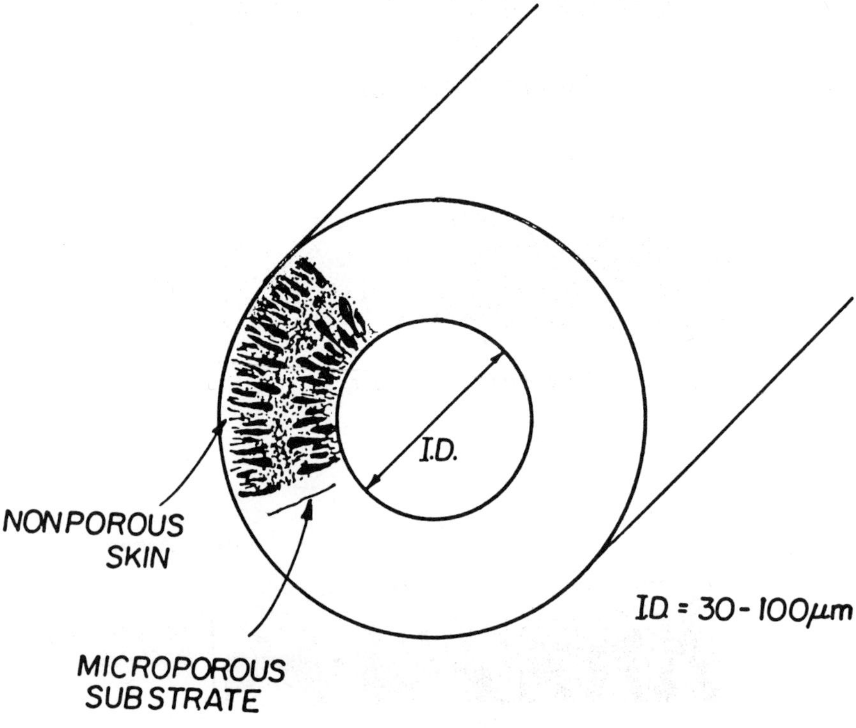

FIGURE 18 Asymmetric hollow fiber

The composite hollow fibers used so successfully by Monsanto Company in its PRISM permeators for the separation of H_2 from various industrial gas mixtures are "skinned" asymmetric hollow fibers of polysulfone coated with a thin film of silicone rubber (~1 μm-thick). The polysulfone skin, which is about 0.1 μm-thick, is the separative element, while the only function of the silicone rubber coating is to plug whatever micropores or discontinuities may be present in the skin (12, 50–52). Polysulfone has a much higher gas-selectivity than silicone rubber. A number of other composite hollow fibers of this type have been studied, for example, polysulfone (1) — polyethylene (2), poly(acrylonitrile (1) — silicone rubber (2), and poly(acrylonitrile)-polyethylene (2), where (1) designates the "skinned" hollow fiber and (2) designates the coating (50–52). No coating is required in cases where the skin is nonporous.

B. Facilitated-Transport Membranes

The rate of permeation of a gas through a membrane can be greatly enhanced by means of an auxilliary mode of transport known as "facilitated" or "carrier-mediated" transport. This mode may be induced in some cases by a reversible chemical reaction between the penetrant gas and a mobile carrier, which is incorporated in the membrane and shuttles between the membrane interfaces (11). Facilitated-transport of a component of a permeating mixture can substantially increase the selectivity of the membrane toward that component.

Conside the permeation of a mixture of components A and B through the facilitated-transport membrane shown diagramatically in Figure 19. such a membrane consists of a microporous support which is impregnated with a non-volatile liquid; the liquid is strongly held in the pores of the membrane by capillary forces. A mobile carrier C which can react reversibly with, say, component A is dissolved in the liquid. The reversible reaction is

$$\mathrm{A} + \mathrm{C} \rightleftharpoons \mathrm{AC},$$

where AC is a complex.

If a pressure (and concentration) gradient of A and B is imposed across the membrane, from left to right, the *forward* reaction will take place at the left interface of the membrane to form AC. A concentration gradient of AC is therefore established in the same direction as that of A. The complex AC then diffuses to the right interface, where the *reverse* reaction takes place to reform A and C. Component A is released from

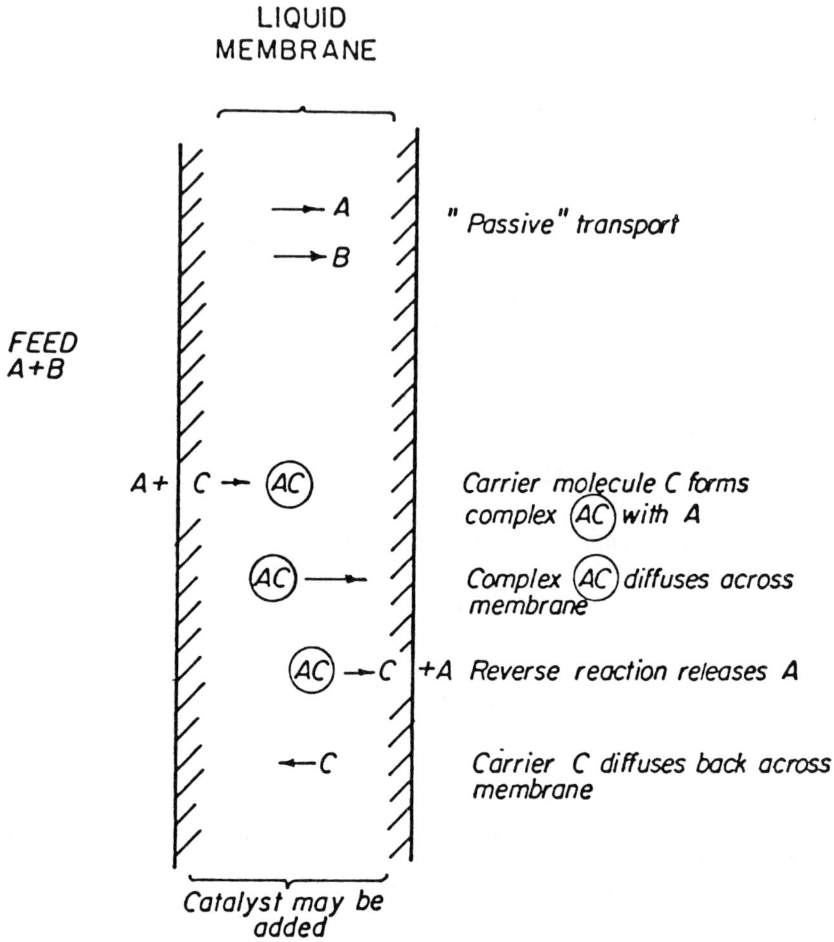

FIGURE 19 Facilitated transport membrane.

solution at the right interface of the membrane. Carrier C, being a nonvolatile molecular or ionic species, diffuses back to the left interface, where it reacts again with A. Thus, C shuttles back and forth between the two membrane interfaces and facilitates the transport of A. The net transport of C is equal to that of AC but in opposite direction.

The selectivity of the membrane is greatly enhanced by the above process, because component A permeates through the membrane both by facilitated transport and by "passive" transport (Fickian diffusion without chemical reaction), whereas component B permeates only by passive transport. The reaction between A and C can be accelerated by addition of a suitable catalyst.

Facilitated transport in synthetic membranes, which mimics a similar process in biological membranes, was studied by Ward (53–57) and by other investigators [see (12) for pertinent references]. It is beyond the scope of this paper to discuss the mechanism of facilitated transport, which has been reviewed elsewhere (58, 59), or the various modifications of this process (12). Facilitated-transport membranes have been developed for O_2, CO_2, CO, H_2S, NO, and olefins, but all suffer from a variety of limitations (12), and therefore have not found as yet any practical applications. The facilitated transport of O_2 is particularly important in air separation, because solid polymer membranes have relatively low separation factors [$^*(O_2/N_2) \lesssim 6$]. A facilitated-transport membrane for O_2 with $\alpha^*(O_2/N_2 = 30$, yielding over 80% O_2, has been developed recently (60). The practical lifetime of this membrane was limited to a few weeks at ambient temperature, or to 3 months at -10°C., because the carrier (a metal-organic complex) oxidized irreversibly and lost gradually its chemical activity.

VI. FUTURE DEVELOPMENTS

The prediction of future developments in membrane separation processes is fraught with danger, because these developments will depend as much on uncertain economic factors, such as the cost of energy, as on technological factors. For example, who would have predicted in 1972 that the cost of oil would quadruple in the interval of one year?

Membrane separation technology has made remarkable progress in the last two decades. The principles of membrane process and equipment design are now well understood, and have been tested on a large scale under field conditions. New membrane processes for the separation of gases will undoubtedly emerge in the coming years. However, the rate of progress of the last 20 years will be sustained only if more highly gas-selective and permeable membranes than available at present can be synthesized. This will require that significant progress be made, in turn, in our understanding of the relationships between the chemical structure of polymers and their gas permeability.

The prior discussion should have made it clear that the selection of a polymer membrane for the separation of a specific gas mixture is based at present on rather primitive considerations. Usually, a process simulation coupled with a preliminary economic analysis first establishes what selectivity and permeability the membrane must exhibit in order to ensure the economic viability of the process. A search is then made in the literature, or in whatever data base is available, for a membrane with the desired properties. If such a membrane indeed exists it is likely to be in the glassy polymer state because, as mentioned earlier, glassy polymers are more gas-selective than rubbery ones (6, 7). However, glassy polymers also exhibit, in general, a low *intrinsic* permeability (i.e. low values of the permeability coefficient $\bar{P}$). Therefore, in order to obtain the desired rate of gas permeation, the selected membrane will have to be used in symmetric or composite form.

It is more likely, however, that no membrane endowed with all the required properties will be found to exist. The desired membrane must then be synthesized by the polymer chemist. The relationships between the chemical structure of polymers and their selectivity and permeability to different gases is not well understood. Therefore, the structure of new membranes with specified permeability characteristics can be found at present only based on some qualitative observations (5, 12, 61), on some semiempirical but useful solubility and diffusivity correlations [e.g. ref. (62)], and on considerable trial-and-error.

Finally, it has been shown earlier that the gas-selectivity of polymer membranes varies inversely, in general, with their intrinsic permeability. Can polymers that exhibit *both* a high selectivity and a high permeability be synthesized? Table III suggests that this might be possible. For example, Japanese investigators (63) have recently developed an acetylenic polymer, poly[1-(trymethylsilyl)-1-propyne)] (PMSP), that exhibits a gas permeability (toward O_2, N_2, CO_2, CH_4, and He) that is about an order of magnitude higher than that of silicone rubber (PDMS). It was previously thought that PDMS was the most gas-permeable polymer because of its large free volume and the flexibility of its Si$-$O bond. By contrast, PMSP is a fairly rigid polymer because its backbone contains alternating double bonds. It is particularly interesting that the O_2/N_2 selectivity of PMSP is about equal to that of PDMS, while the CO_2/CH_4 selectivity is higher.

Table III also shows that the glassy polymers poly(methylpentene) and poly(oxydimethyl phenylene) exhibit a higher O_2 permeability at 25°C than rubbery polyethylene and butyl rubber, and as high a permeability as that of Buna S. At the same time, the O_2/N_2 selectivity of the two glassy polymers is higher than that of the three rubbery polymers. Clearly, much remains to be learned in this field.

Significant efforts are being made at present in a number of academic and industrial laboratories to advance the materials science of gas

TABLE III

Selected gas permeabilities and separation factors of rubbery and glassy polymers

Polymer	Permeability coefficients, $\bar{p} \times 10^{10}$ cm^3(STP)cm/(s·cm^2·cm Hg)				Ideal separation factor, $\alpha^*(A/B)$	
	O_2	N_2	CO_2	CH_4	O_2/N_2	CO_2/CH_4
Rubbery Polymers						
Poly(1-trimethylsilyl-1-propyne)	4200	2100	18000	4200	2.0	4.3
Poly(dimethylsiloxane)	500	238	2700	800	2.1	3.4
Natural rubber	24	8.1	131	30	3.0	4.4
Buna-S	17.1	6.32	124	21	2.7	5.9
Polyethylene($\rho = 0.914$)	2.9	0.97	12.6	2.9	3.0	4.3
Butyl rubber	1.3	0.33	5.18	0.78	3.9	6.6
Glassy Polymers						
Poly(methylpentene)	18.9	4.5	—	—	4.1	—
Poly(oxydimethyl phenylene)	15.8	3.81	75.7	—	4.1	—
Polystyrene	7.47*	2.55*	23.3*	2.73*	2.9	8.5*
Poly(bis a carbonate) (Lexan)	1.40	0.300	10*	0.5*	4.7	20*
Polysulfone**	—	0.19	6.4	0.23	—	28**
Poly(vinyl fluoride)	0.020	0.0042	—	—	4.8	—

$\alpha^*(A/B) \equiv \bar{p}_A/\bar{p}_B$; temperature = 25°C; *35°C; **35°C.

separation membranes. These efforts should reself in important new developments in membrane technology for gas separations.

ADDENDUM: A detailed study of recycle permeators has recently been reported by McCandles (63).

REFERENCES

1. T. Graham, Phil. Mag., *32*, 401 (1866)
2. G. Parkinson, Chemical Engineering, April 16 (1984), p. 14
3. S. Loeb and S. Sourirajan, "Sea Water Demineralization by Means of a Semipermeable Membrane," University of California (Los Angeles), Dept. Eng. Repts., No. 60–60 (1961).
4. C.E. Rogers, "Solubility and Diffusivity", Chapt. 6 in "Physics and Chemistry of the Organic Solid State", D. Fox, M.M. Labes, and A. Weissberger, Eds., Interscience Publishers, New York, 1965, pp. 510–635.
5. S.A. Stern, "Gas Permeation Processes", Chapt. XIII in "Industrial Processing with Membranes", R.E. Lacey and S. Loeb, Eds., Wiley-Interscience, New York, 1972, pp. 279–339.
6. S.A. Stern and H.L. Frisch, Ann. Rev. Mater. Sci., *11*, 523 (1981).
7. H.L. Frisch and S.A. Stern, "Diffusion of Small Molecules in Polymers", CRC Crit. Rev. in Solid State and Mat. Sci., *11*, No. 2, 123–187 (1983), CRC Press, Boca Raton. FL
8. J. Crank, "The Mathematics of Diffusion", 2nd Ed., Clarendon Press, Oxford, 1975
9. W. Jost, "Diffusion in Solids, Liquids, Gases", Academic Press, New York, 1952
10. S.-T. Hwang and K. Kammermeyer, "Membranes in Separations", Wiley-Interscience, New York, 1975
11. S.A. Stern, "The Separation of Gases by Selective Permeation", Chapt. 9 in "Membrane Separation Processes", P. Meares, Ed., Elsevier Scientific, New York, 1976, pp. 296–326
12. S.L. Matson, J. Lopez, and J.A. Quinn, Chem. Eng. Sci., *38*, 503 (1983)
13. S. Weller and W.A. Steiner, J. Appl. Phys., *21*, 279 (1950)
14. S. Weller and W.A. Steiner, Chem. Eng. Progr., *46*, 585 (1950)
15. R.W. Naylor and P.O. Backer, AIChE J., *1*, 95 (1955)
16. M.E. Breuer and K. Kammermeyer, Separation Sci., *2*, 319 (1967)
17. S.A. Stern and W.P. Walawender, Separation Sci., *4*, 129 (1969)
18. W.P. Walawender and S.A. Stern, Separation Sci., *7*, 553 (1972)
19. C.T. Blaisdell and K. Kammermeyer, Chem. Eng. Sci., *28*, 1249 (1973)
20. C.Y. Pan and H.W. Habgood, Ind. Eng. Chem. Fundam., *13*, 323 (1974)
21. C.Y. Pan and H.W. Habgood, Can. J. Chem. Eng., *56*, 197 (1978): ibid., *56*, 210 (1978)
22. S.A. Stern and S.-C. Wang, J. Membrane Sci., *4*, 141 (1978)
23. S.A. Stern and F.J. Onorato, and C. Libove, AIChE J., *23*, 567 (1977)
24. S.A. Stern and S.M. Leone, AIChE J., *26*, 881 (1980)
25. S.A. Stern and S.-C. Wang, AIChE J., *26*, 891 (1980)
26. S.A. Stern. J.E. Perrin, and E.J. Naimon, J. Membrane Sci., *20*, 25 (1984)
27. W.C. Pfefferle, U.S. Patent No. 3,144,313, August 11, 1964
28. S.-T. Hwang and J.M. Thorman, AIChE J., *26*, 558 (1980)
29. S. Kimura, T. Nimura, T. Mijauchi, and M. Ohno, Radiochem. Radioanal. Lett., *13*, 349 (1973)
30. M. Ohno, T. Morisue, O. Ozaki, H. Heki, and T. Miyauchi, Radiochem. Radioanal. Lett., *27*, 299 (1976)
31. M. Ohno, O. Ozaki, and H. Sato, J. Nucl. Sci. Technol. Japan, *14*, 589 (1977)
32. M. Ohno, T. Morisue, O. Ozaki, and T. Miyauchi, J. Nucl. Sci. Technol. Japan, *15*, 376 (1978)

33. M. Ohno, H. Heki, O. Ozaki, and T. Miyauchi, J. Nucl. Sci. Technol. Japan, *15*, 668 (1978)
34. K.K. Sirkar, Sepn. Sci. Technol., *15*, 1091 (1980)
35. A. Sengupta and K.K. Sirkar, J. Membrane Sci., *21*, 73 (1984)
36. J.E. Perrin and S.A. Stern, to be published in AIChE J.
37. R.F. Kesting, "Synthetic Polymeric Membranes", McGraw-Hill Book Co., New York 1971
38. H.K. Londsale and H.E. Podall, "Reverse Osmosis Membrane Research", Plenum Press, New York, 1972
39. H. Strathmann, "Trennung von Molekularen Mischungen mit Hilfe Synthetischer Membranen", Steinkopff Verlag, Darmstadt, 1979
40. W. Pusch, Angew. Chem., *21*, 660 (1982)
41. P. Aptel, Informations Chimie, No. 213, 189, May 1981
42. K.D. Vos and F.O. Burris, Ind. Eng. Chem., Prod. Res. Develop., *8*, 84 (1969)
43. J.P. Agarwal and S. Sourirajan, J. Appl. Polym. Sci., *14*, 1303 (1970)
44. W.M. MacDonald and C.Y. Pan, U.S. Patent 3,842,515 (1974)
45. R.L. Riley, H.K. Lonsdale, C.R. Lyons, and V. Merten, J. Appl. Polym. Sci., *11*, 2143 (1967)
46. L.T. Rozelle, J.E. Cadotte, and D.J. McClure, "Membranes from Cellulose and Cellulose Derivatives", A.F. Turbak, Ed., J. Wiley & Sons, New York, 1970, pp. 61–71
47. W.J. Ward, III, W.R. Browall, and R.M. Salemme, J. Membrane Sci., *1*, 99 (1976)
48. W.J. Ward, III, U.S. Patent 4,279,855 (1981)
49. S.G. Kimura, R.G. Lavigne, and W.R. Browall, U.S. Patents 4,132,824 (1979) and 4,192,842 (1980)
50. J.M.S. Henis and M.K. Tripodi, Separation Sci. Technol., *15*, 1059 (1980)
51. J.M.S. Henis and M.K. Tripodi, U.S. Patent 4,230,463 (1980)
52. J.M.S. Henis and M.K. Tripodi, J. Membrane Sci., *8*, 233 (1981)
53. W.J. Ward, III, and W.L. Robb, Science, *156*. 1481 (1967); ibid. *227*, 162 (1970)
54. W.J. Ward, III, in "Recent Developments in Separation Science", Vol. 1, N.N. Li, Ed., CRC Press, New York, 1972, pp. 153–161
55. W.J. Ward, III, U.S. Patents 3,780,496 (1973), 3,819,806 (1974), and 4,115,514 (1978)
56. S.L. Matson, C.S. Herrick, and W.J. Ward, III, Ind. Eng. Chem. Proc Des. Dev., *16*, 370 (1977)
57. S.G. Kimura, S.L. Matson, and W.J. Ward, III, "Recent Developments in Separation Science", N.N. Li, Ed., Vol. 5, CRC Press, Boca Raton, FL, 1979, p. 11
58. K.A. Smith, J.H. Meldon, and C.K. Colton, AIChE J., *19*, 102 (1973).
59. J.S. Schultz, J.D. Goddard, and S.R. Suchdeo, AIChE J., *20*, 417 (1974); ibid., *20*, 625 (1974)
60. R.W. Baker, I.C. Roman, K.L. Smith, and H.K. Lonsdale, paper presented at 9th Energy Technology Conf., Washington, DC, February 17, 1982
61. H.H. Hoehn, in "Materials Science of Synthetic Membranes", ACS Symposium Series 269, American Chemical Scoiety, Washington, D.C., 1985, pp. 81–98
62. R.T. Chern, W.J. Koros, H.B. Hopfenberg, and V.T. Stannett, in "Materials Science of Synthetic Membranes", ACS Symposium Series 269, American Chemical Society, Washington, D.C., 1985, pp. 25–46
63. T. Masuda, E. Isobe, and T. Higashimura, J. Am. Chem. Soc., *105*, 7473 (1983)
64. F.P. McCandless, J. Membrane Sci., *24*, 15 (1985).

Transport Through Membranes

C. E. ROGERS AND ALEXANDER SFIRAKIS*

Department of Macromolecular Science Case Western Reserve University Cleveland, Ohio 44106

Membrane materials can be classified as either microporous, homogeneous (solution-diffusion mechanism), or electrically charged. Regardless of the classification, transport of penetrant through a membrane can be considered as the product of functions of the amount of penetrant in the membrane (an effective solubility), average penetrant mobility (an effective diffusivity), and the driving force. These functions are very sensitive to the relative physiochemical natures of the membrane material and the penetrant molecule, especially in terms of interactions which affect the mode of sorption and mobility of different penetrant species. The nature and effects of several factors affecting transport behavior are discussed with emphasis on homogeneous membrane systems. Topics include the different effects of water and alcohol clustering on diffusion and dynamic mechanical properties in Nylon 6 and the effects on alcohol sorption and diffusion behavior of going from a rubbery to a glassy state with changing copolymer composition. These topics illustrate effects of component interactions as they may affect sorption, transport, and other properties.

INTRODUCTION

The use of polymeric materials as protective barrier films and as permselective membranes for separation processes has undergone astounding growth over the last decade. The scope of applications, features of various film and membrane systems, and fundamental aspects of the nature and mechanisms of sorption and transport have been discussed in a number of recent reviews (e.g., 1–8).

The selection or development of polymeric materials for use in applications with stringent design specifications relating to their solution and transport properties, as well as other properties such as cohesive and/or adhesive strength, effects of environmental exposure, etc., requires knowledge and appreciation of the many chemical and physical factors which affect these properties. In the case of gaseous penetrants, our

* Present Address: Resinas Sinteticas, S.A. Apartado Postal No. 175 Código Postal 58000 Morelia, Michoacán MEXICO

present level of knowledge seems fairly adequate for interpreting and predicting the behavior of simple systems. As the size of the penetrant increases and/or as component interactions increase, our understanding of the resultant solution and transport behavior becomes less clear. The situation is becoming even more complex with the increasing use of multicomponent, often multiphase, polymer mixtures as films and membranes. It is apparent that there must be a considerable research effort made to gain a better understanding of the behavior of these more complex polymer-penetrant systems to guide design and development procedures.

In this paper we will consider aspects of two factors which affect solution and transport behavior in polymeric materials. The first is an aspect of the factor we may term "the mode of sorption", defined as the manner by which penetrant molecules are accommodated within the polymer matrix (random dispersal, solvation shells, clustering, etc.) as it affects concurrent diffusion and other properties.

The second is an aspect of the factor relating the change in solution and transport behavior of a given penetrant in a given polymer on changes in polymer "morphology" as the temperature of the experiment is taken past the system's glass temperature. Morphology, in this case, is used in its widest sense to describe the physical nature of materials, be they semicrystalline or amorphous. For glassy amorphous polymers it relates to the amount and distribution of excess volume or "frozen holes" or related structural features of the glass state.

Factors Affecting Sorption and Transport

In a binary mixture of polymer and penetrant, the driving force causing diffusion of penetrant is proportional to the negative of its chemical potential gradient $-(\partial\mu/\partial x)$ at any given point. The total force at a given point therefore is proportional to $-c(\partial\mu/\partial x)$ and the flux due to pure diffusion (through a properly defined reference plane to account for mass flow, etc. (8)) can be written as:

$$J = -(m'RT/N)(\partial \ell na'/\partial \ell nc')(\partial c'/\partial c)(\partial c/\partial C)(\partial C)\partial x) \qquad (1)$$

where N is Avogadro's number, a is the thermodynamic activity ($\partial\mu = RT\partial \ell na$), c is the total sorbed concentration in equilibrium with the ambient concentration C with the overall solubility coefficient $S = (\partial c/\partial C)$, and m is the intrinsic mobility of a penetrant molecule for unit force.

The quantity $(\partial c'/\partial c)$ is a measure of the distribution of sorbed penetrant molecules in different modes of sorption which affects the mobilities of the sorbed penetrant. In the simplest case, the dual sorption model can be imposed with the concentration of free-to-diffuse molecules denoted by c′ with an average mobility m′. The quantity $(\partial \ell na'/\partial \ell nc')$ is a correction for ideality of solution behavior; for an ideal solution it is unity. It also can be expressed as $(1 + \partial \ell n\gamma/\partial \ell nc)$, where γ is the activity coefficient defined as $\gamma = a/c$ (or $\gamma' = a'/c'$). If, for any reason, these terms become negative in sign, diffusion may occur against the concentration gradient as has been observed in biological systems (e.g., 11) and in synthetic polymer membranes possessing a gradient of chemical composition (e.g., 12), or temperature, or stress, or other variable which affects diffusion in terms of irreversible thermodynamics.

The "true" diffusion coefficient

$$D' = (m'RT/N)(\partial \ell na'/\partial \ell nc'), \tag{2}$$

which is a measure of the average mobility of mobile penetrant molecules, is related to experimentally measured values, D, from the usual definition of flux, $J = DS(\partial C/\partial x)$, as:

$$D = D'(\partial c'/\partial c) \tag{3}$$

The measured D is equal to D′ only when all sorbed penetrant molecules are mobile in the sense of the dual sorption model. More generally, the above argument can be reconsidered in terms that $(\partial c'/|c)$ is a function describing the distribution of penetrant molecules in modes of sorption with characteristic mobilities represented as a D′ function (12). This allows the system to be considered as a mixture with a distribution of component interactions, rather than as a mixture of penetrant sorbed by two different mechanisms (e.g., Langmuir + Henry's Law models).

The concentration (and temperature) dependence of the diffusion coefficient for a given system is seen to arise from three sources: the concentration dependence of the mobility, the variation of the concentration of different mobile species with total concentration, and a concentration dependence attributed to the thermodynamic ideality of the mixture. The influence of these sources on the diffusion process will depend on the nature of interactions between the components, the structural characteristics of the polymeric matrix and any changes induced by the presence of penetrant, as well as the experimental conditions of time (relaxation processes), temperature, and ambient penetrant concentration. It is anticipated that other properties of the polymer, such as mechanical and dynamic-mechanical properties, will be affected by the presence of a penetrant sorbed in a distribution of sorption modes or inducing changes in polymer structure.

Sorption Mode Effects on Transport and Dynamic Mechanical Properties

The sorption isotherms of water, methanol, and ethanol at 35°C in films of Nylon 6 of 1.25 mils thickness were determined using a quartz helix vacuum microbalance and a Cahn RC recording electrobalance (13). The two alcohols followed the same volume fraction, ϕ_1, versus vapor activity, a_1, isotherm ($\phi_1 = 0.15$ for $a_1 = 0.9$). The isotherm for water was lower ($\phi_1 = 0.08$ for $a_1 = 0.9$). Both curves were convex to the vapor activity axis with a small BET-type concavity indicated at very low a_1 values. The BET equation in its linear form (14) fit the data only at the lower relative vapor pressures. A Flory-Huggins type of isotherm equation, modified to account for the crystalline volume excluded from the sorption process and the restraint on swelling due to intercrystallite tie molecules (15), gave values of the interaction parameter, χ_1, that, for water, were 1.35 below $C_1 = 0.1$ moles of water/100 grams of polymer and 1.5 above that concentration. The χ_1 values for the alcohols increased with increasing concentration from, for methanol, 0.1 at $C_1 = 0\,m/100\,g$ to 1.0 at $C_1 = 0.37\,m/100\,g$ and, for ethanol, 0.2 at $C_1 = 0\,m/100\,g$ to 1.0 at $C_1 = 0.24\,m/100\,g$.

Analysis of the sorption isotherms was made using the clustering function equation (16):

$$G_{11}/v_1 = -(1 - \phi_1)[d(a_1/\phi_1)/da_1]_{p,T^{-1}} \tag{4}$$

where v_1 is the partial molecular volume and G_{11} is the cluster integral. The cluster function, G_{11}/v_1, is a measure of the clustering tendency of penetrant molecules in the mixture. A value of -1 is characteristic of an ideal, random solution mixture. Values less than -1 indicate that sorption is occurring on specific polymer sites so that the isotherm is concave to the concentration axis (e.g., Langmuir type). Values greater than -1 indicate that penetrant-penetrant molecule interactions are preferred leading to formation of penetrant clusters within the polymeric matrix.

The plot in Figure 1 of the cluster function versus sorbed concentration indicates a transition from specific site sorption to clustering at concentrations (moles/100 grams) of about 0.1 for water and 0.17 for the alcohols. Water, being more polar, shows an apparent higher clustering tendency than do the alcohols. The higher transition concentration for the alcohols, approximately twice that for water, might be related to the

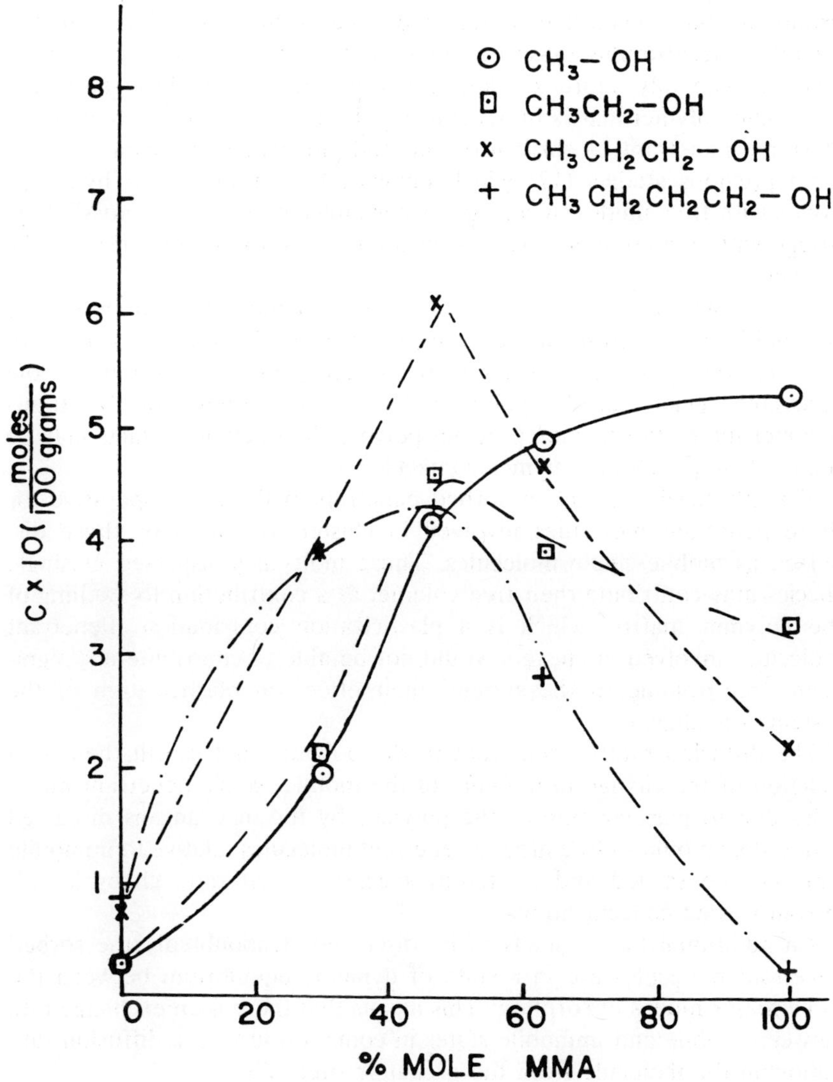

FIGURE 1 Clustering function vs. sorbed concentration in Nylon 6 at 35°C.

relative hydrogen bonding capacity difference between water and alcohols.

These results indicate that there are at least three modes of sorption of water and alcohols in this polymer: specific site sorption, presumably a hydrogen bonding interaction between the penetrant and free amide groups on the polymer; penetrant molecule cluster formation at higher overall concentrations of sorbed penetrant; and single penetrant molecules presumably more or less randomly dispersed throughout the accessible polymer matrix in dynamic equilibrium with the site-bound and clustered penetrant. A forth sorption mode for water has been reported from previous studies (17) which involves formation of two hydrogen bonds with free amides by a single water molecule to form a crosslinking bridge with a consequent increase in the α-transition temperature of the polymer.

The formation of specific site penetrant-polymer interactions shields the amide groups from participating in intra- or interchain interactions thus removing some restriction on chain-segmental motion in the accessible regions of the polymer. Due to the affinity of penetrant-polymer interactions, some polymer-polymer interaction contacts may be replaced by penetrant-polymer contacts.

This plasticizing effect of sorbed penetrant will not be operative for those penetrant molecules involved in cluster formation or those dispersed as mobile single molecules. Those molecules dispersed as single species may contribute their free volume, as a contribution to swelling of the polymer matrix, which is a plasticization contribution. Penetrant molecules involved in clusters would not be able to contribute any significant free volume to the system; their effect on plasticization of the system is negligible.

The diffusion rate of penetrant in the polymer matrix will then be a function of the change in mobility of the mobile, single penetrant molecules due to plasticization of the polymer by the mechanisms discussed and to the amount of free mobile penetrant molecules relative to immobile (specific site sorbed and clustered) species as that ratio changes with overall sorbed concentration.

An additional factor needs to be discussed. Undoubtedly the sorbed penetrant molecules are in a state of dynamic equilibrium between the three major modes of sorption. This means that there is an exchange rate between mobile and immobile states in competition with a diffusion rate removing the molecule from the exchange site:

$$\text{(immobile)} \underset{k}{\overset{k'}{\rightleftharpoons}} \text{(mobile)} \overset{D'}{\rightarrow} \text{(diffused)}$$

where $k'/k = K$, the equilibrium coefficient characterizing the ratio c'/c. If the absolute values of k' and k are small relative to the correlation time scale for diffusion characterized by D', then a given penetrant molecule resides in the immobilized state for a time greater than that of the diffusive jump frequency. Consequently, $c' < c$ and $D < D'$.

When k′ and k are comparable to or greater than the diffusion correlation time scale, the molecules are exchanging so rapidly that there is little effect on the diffusion process due to immobilization. This effect of exchange rate relative to diffusion rate has been treated (18, 19) by assigning a diffusion coefficient characterizing the "partial" mobility of the "immobilized" penetrant in addition to the diffusion coefficent of mobile penetrant.

In addition, the exchange rate must be considered in terms of effects of sorbed penetrant on other physical properties. Each property will have a characteristic time scale or frequency which, in comparison to the exchange rate or frequency, will determine the effect of penetrant localization on the particular property.

These considerations are illustrated by the diffusion data shown in Figure 2 and by the dynamic-mechanical data shown in Figure 3. In the case of diffusion, we note that the concentration dependence is much less pronounced for water than for the alcohols. The dependence of the diffusion coefficient for water on concentration is linear, on the log scale, over the entire concentration range. The dependence of logD for the alcohols on concentration is very pronounced at low concentrations, but then level off to about that of water at higher concentrations. The concentration at which the alcohol data level off corresponds closely to the advent of cluster formation, about 0.2 moles/100 g of polymer.

In Figure 3, the data for the change in intensity of the γ-relaxation peak as a function of sorbed water concentration shows a leveling off at a concentration corresponding to the apparent onset of water clustering as shown in Figure 1. The data for the alcohols shows a similar leveling off at the concentration at which both the onset of clustering occurs and the diffusion coefficient data level off.

Consideration of the mechanism ascribed to the γ-relaxation process affords some further insight into the effects of sorbed penetrant on that behavior. It has been proposed (e.g., 20) that the process involves methylene units which are restrained by hydrogen-bonded polar groups. The sorption of water and alcohols, which may undergo more-or-less specific hydrogen-bonding interactions with those polar groups, decreases the number of units participating in the relaxation process. This decreases the relaxation peak intensity with little change in peak temperature. As penetrant concentration increases, penetrant clustering becomes dominant. Since penetrant molecules involved in cluster formation do not further affect the number of units participating in the γ-relaxation process the peak intensity levels off.

These penetrant-polar group interactions are on a different time scale than the diffusion process and so the effect of water clustering to limit the free water available for polar group interaction is of significance. The

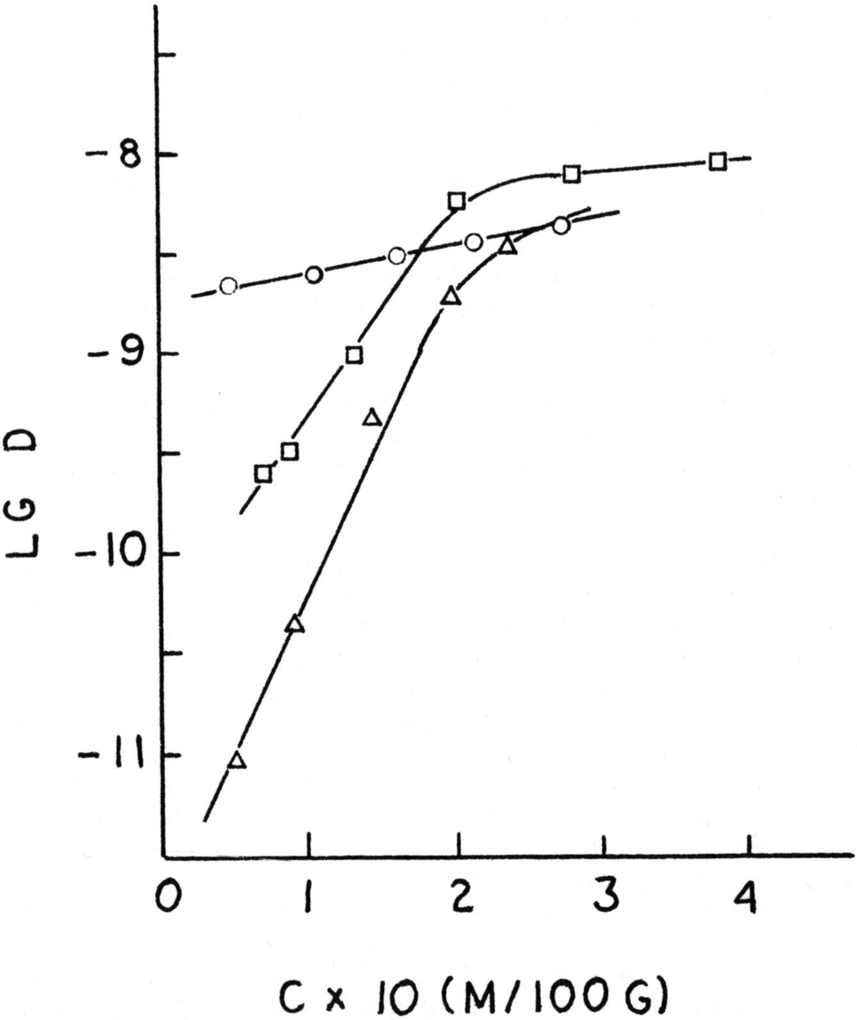

FIGURE 2 Dependence of diffusion coefficients of o water, □ methanol, and △ ethanol on sorbed concentration in Nylon 6 at 35°C.

effect of clustering of the alcohols is of significance for both diffusion and relaxation processes. These differences between water and alcohols are a consequence of their molecular size as it affects mobilities (and exchange rates) and their inherent polarities (or hydrogen-bonding functionalities).

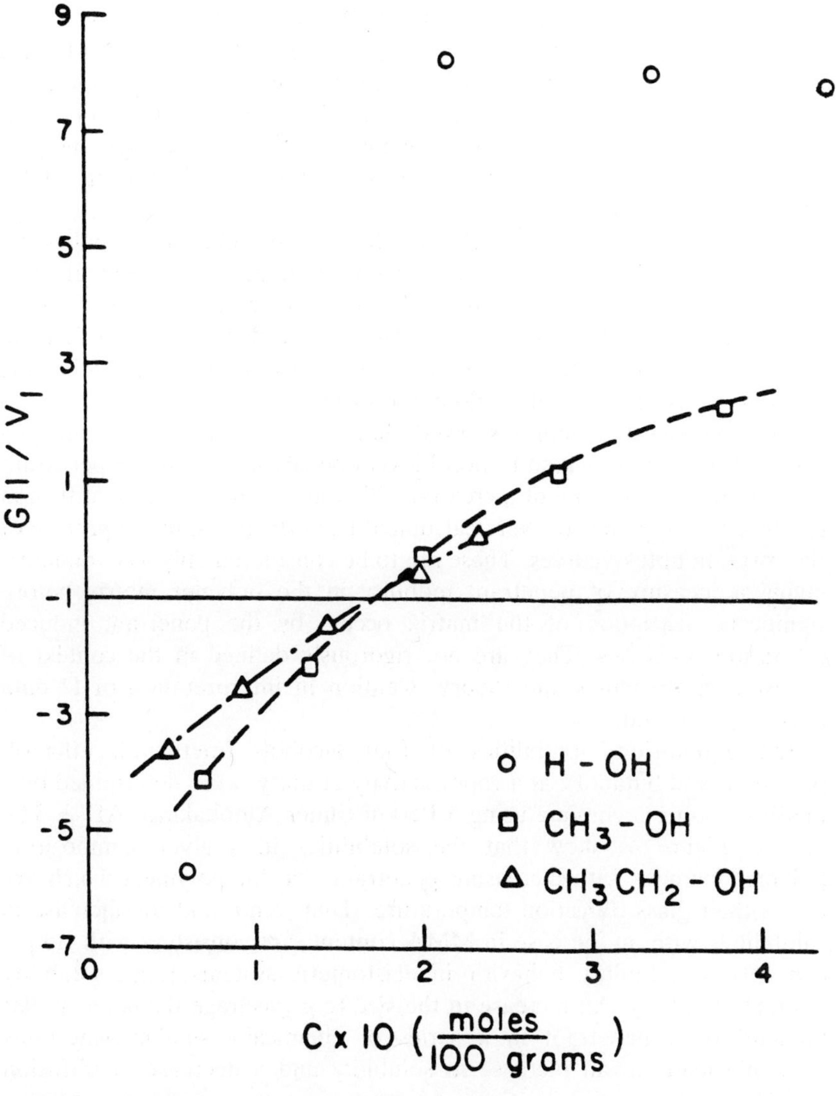

FIGURE 3 Dependence of intensity of the γ-relaxation peak in Nylon 6 on sorbed o water, □ methanol, and △ ethanol.

Sorption and Diffusion in Rubbery and Glassy Polymers

The effects of polymer composition and penetrant molecular size on the solubility and diffusivity of alcohol vapors in a series of well characterized isoprene (IP)-methyl methacrylate (MMA) copolymers have been investigated at room temperature (21). At 55 mole percent MMA the copolymer undergoes a rubber to glass transition at the temperature of the experiment. It was of interest, in this study, to determine how solubility and diffusivity behavior changed as the polymer systems progressed from the elastomeric to glassy states with increasing MMA composition.

Copolymers of IP and MMA and the corresponding homopolymers, PIP and PMMA, were prepared by emulsion polymerization at 50°C using a modified GR-S formulation (22). Characterization by GPC, DSC, NMR, mechanical property, and density measurements showed that the polymers were amorphous with copolymer physical proporties following relations derived for ideal random copolymers.

Sorption kinetics studies showed that the sorption process changed progressively from Fickian to non-Fickian behavior both as the penetrant size and the MMA content increased. What are termed here as "diffusion coefficients", denoted $\tilde{D}$, were calculated from the initial linear portion of the sorption uptake curves. These are to be considered only as parameters giving a measure of penetrant mobility in the polymer matrix before significant disruption of the matrix occurs by the penetrant induced relaxation processes. They are not rigorously defined in the context of diffusion mathematics and theory. Caution in interpretation of $\tilde{D}$ data must be exercised.

The equilibrium solubilities of four alcohols (methanol, ethanol, propanol, and butanol), at a vapor activity of unity, were determined by a modified bottle technique using a Perkin-Elmer Autobalance AD-2. The results, Figure 4, show that the solubilities in a given composition polymer increase with increasing penetrant size for polymers which are above their glass transition temperature. That trend, and the increase in solubilities with an increase in MMA content, are consistent with simple concepts of solubility behavior in elastomeric systems (e.g., solubility parameter) (7, 8). An increase in the size (e.g., average diameter, molar volume) of a penetrant in a series of chemically similar penetrants generally leads to an increase in solubility and a decrease in diffusion coefficients, Figure 5. Likewise, an increase in polar content in a polymer series increases the solubility of a polar (and hydrogen bonding) penetrant.

Sorption and diffusion processes in the glassy state generally show a even greater dependence on the size (and shape (21)) of the penetrant molecule than in the rubbery state (7, 8). This can be ascribed to a change in the process of mixing from one of site generation above T_g to the effective filling of preexisting sites governed by the excess volume distribution in the more rigid glassy polymer. This is illustrated in Figure 4

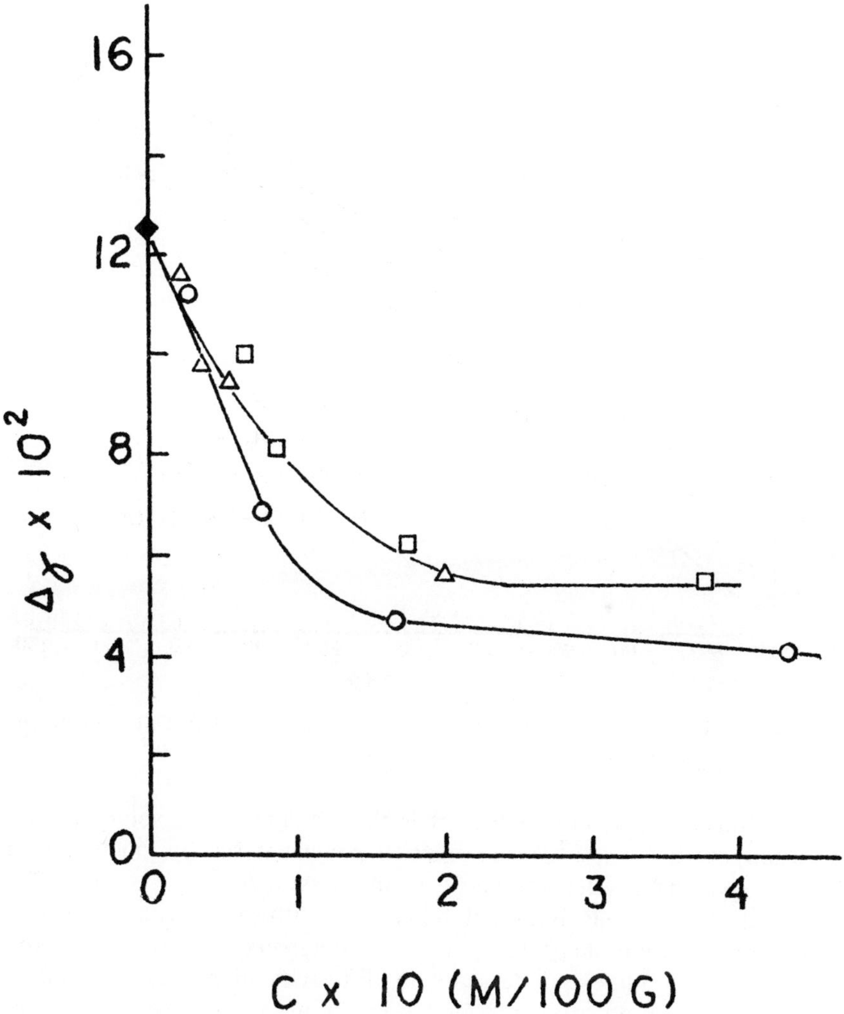

FIGURE 4 Concentration of sorbed alcohols vs. polymer composition.

where the order of solubility magnitudes, in terms of penetrant size, is reversed as the polymer system goes from rubbery to glassy for MMA contents greater than about 55%. This suggests that the distribution of excess volume provides fewer sorption sites for large molecules than for small. A corresponding change in $\tilde{D}$ also is noted, Figure 5, resulting in larger differences in $\tilde{D}$ between penetrants in glassy systems than in corresponding rubbery systems.

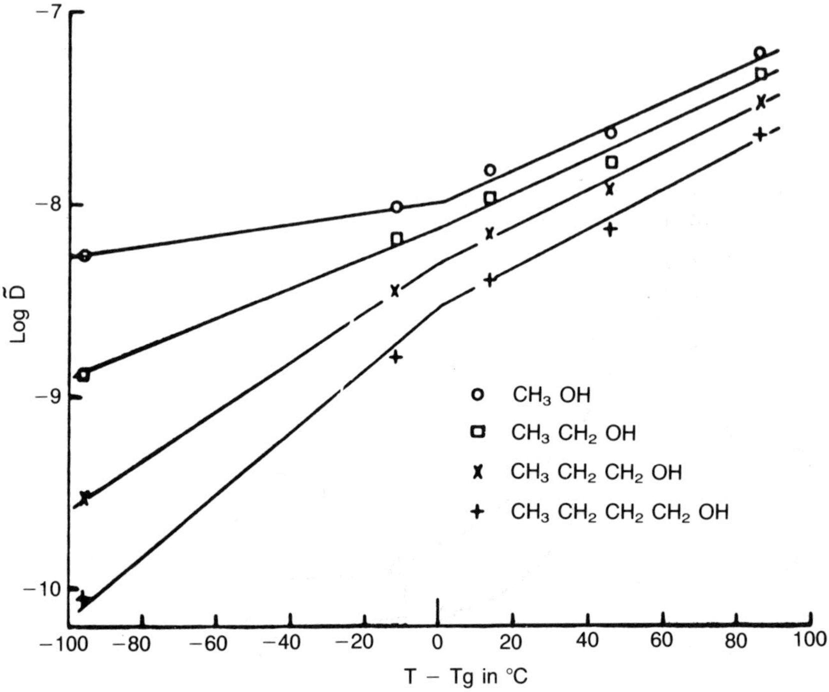

FIGURE 5 "Diffusion coefficients" versus $T-T_g$ for the series of IP-MMA copolymers and corresponding homopolymers.

Since the permeability coefficient is the product of the solubility and diffusion coefficients, its variation with penetrant size is much less in rubbery polymer systems than it is in glassy polymer systems. The observed decrease in both solubility and diffusivity with increasing penetrant size (and shape (21)) in glassy polymers, compared with the increase in solubility and decrease in diffusivity in rubbery polymers, means that glassy polymer systems are inherently more permselective, in the absence of complicating plasticization and relaxation effects. This is a partial basis for the use of glassy polymers as prospective membrane systems for separation of gaseous mixtures,

Conclusion

It is seen that there can be considerable interaction between sorbed diffusing penetrant molecules and the membrane polymeric material. Differing modes of sorption, as they change with concentration, time, and position within the membrane, may significantly affect plasticization processes, the localization/mobility of penetrant, and other properties. The degree of chain segmental mobility of the polymer is a major factor governing not only the diffusion process but also the solubility process insofar as it controls sorption site availability in a glassy polymer in addition to the thermodynamic factors controlling solution behavior in any system. Again, it is evident that more research must be done to clarify the nature of these penetrant-polymer interactions and their effects on system structure and properties.

These investigations were supported by grants from the National Science Foundation (ENG 77-20037).

REFERENCES

1. H.K. Lonsdale, J. Membrane Sci., *10*, 81 (1982).
2. D.R. Paul and G. Morel, Kirk-Othmer Encyc. of Chem. Tech., 3rd Ed., *15*, 92 (1981).
3. H. Strathmann, J. Membrane Sci., *9*, 121 (1981).
4. T. Parrett, Technology, *3*, 16 (Mar/Apr 1982).
5. S.-T. Hwang and K. Kammermeyer, *Membranes in Separations*, John Wiley and Sons, New York, 1975.
6. P. Meares, ed., *Membrane Separation Processes*, Elsevier Sci. Publ. Co., Amsterdam, 1976.
7. J. Comyn, ed., *Polymer Permeability*, Elsevier Appl. Sci. Publ., London, 1985.
8. J. Crank, *The Mathematics of Diffusion*, 2nd. Ed., Oxford Univ. Press, London, 1975.
9. P. Meares, J. Polym. Sci., *27*, 405 (1958); Trans. Faraday Soc., *54*, 40 (1958).
10. G.S. Park, Trans. Faraday Soc., *48*, 11 (1952).
11. L.F. Nims, Adv. Bio. Med. Phys., *11*, 159 (1967).
12. C.E. Rogers, J. Membrane Sci., submitted.
13. A. Sfirakis and C.E. Rogers, Polym. Engr. Sci., *20*, 294 (1980).
14. A.W. Adamson, *Physical Chemistry of Surfaces*, 4th Ed., John Wiley & Sons, New York, 1982, Chap. XVI.
15. C.E. Rogers, V. Stannett and M. Szwarc, J. Phys. Chem., *63*, 1406 (1959).
16. B.H. Zimm and J.L. Lundberg, J. Phys. Chem., *60*, 425 (1956).
17. C.E. Rogers, J.R. Semancik and S. Kapur, Polym. Sci. Technol., *1*, 297 (1973).
18. J.H. Petropoulas, J. Polym. Sci., A2, *8*, 1797 (1970).
19. D.R. Paul and W.J. Koros, J. Polym. Sci., Phys. Ed., *14*, 675 (1976).
20. Y.S. Papir, S. Kapur, C.E. Rogers and E. Baer, J. Polym. Sci., A2, *10*, 1305 (1972).
21. A. Sfirakis and C.E. Rogers, Polym. Engr. Sci., *21*, 542 (1981).
22. C.S. Marvel, J.L.R. Williams and H.E. Baungarten, J. Polym. Sci., *4*, 583 (1948).

"Membrane Systems"[1]

SEYMOUR S. KREMEN

Vice President, Hydranautics, 8388 Vickers Street — Suite 206, San Diego, CA 92111, U.S.A.

The useful work and economic value which can be obtained by use of membranes for fluid processing is strongly dependent on the physical device (configuration) in which the membrane is housed, in combination with the collateral processes (pretreatment and post-treatment) which are essential to produce a practical useful result.

During the past twenty years numerous configurations have been proposed and tested for uses ranging from water purification and reclamation to dewatering (without heat) to produce a high value concentrate. Marketplace economics have acted to produce a situation which may be summarized as follows: where the value of the purified water produced is expressed in dollars per kilogallon or acrefoot, the spiral wound and hollow fine fiber configurations have become the sole choices, with plant capacities now commonly in the mega-gallon per day size range; where the value of the product, be it permeate or concentrate, is expressed in dollars per gallon or per pound, other configurations, usually variants of the tubular or plate and frame designs, continue to be employed, generally in systems whose capacities are less than 50,000 gal. per day.

These marketplace divisions also reflect large differences in cleanability, pretreatment, acceptability of shutdowns for sanitizing, etc. A number of configurations and membrane processing systems are described and reviewed in terms of their separations process requirement objectives, applicability, and economic considerations.

I. INTRODUCTION

Membrane development usually commences with laboratory preparations, often not much larger in size than a playing card. If the membrane is to be accepted and used in the commercial world with any degree of lasting success, there are a number of considerations which have to be addressed.

These include but are not limited to:

a) The methods and equipment which will ensure that the membrane can be produced repetitively to meet resonable performance limits under expected service conditions;

[1] This paper was presented at the "International Symposium — Synthetic Membranes", Michigan Molecular Institute, Midland, MI, USA, August 20–24, 1984.

b) The configuration in which the membrane is to be packaged and used; in other words, what sort of device will be required to house and support the membrane, etc.?

c) Defining the major application areas and process conditions necessary to the successful utilization of the combination of membrane and chosen device(s).

These reduction-to-practice considerations have received a good deal of attention and effort over the past twenty years. Essentially, four basic configurations (plus numerous variants within each) have been proposed and developed. The four principal reverse osmosis and ultrafiltration membrane configurations may be listed as follows:

aa — Plate and Frame
bb — Tubular
cc — Hollow Fiber
dd — Spiral Wound

Over the years, the particular applications areas, for which each configuration was best suited in terms of technical performance at costs consistent with the value of the product, became generally accepted.

As in most other areas of technical and applications efforts, the objectives sometimes seemed mutually inconsistent, necessitating many compromises and trade-offs. Nevertheless, all approaches to configuration and process suitability start from a common conceptual base. This is illustrated by the simple schematic in Figure 1, which is applicable to all RO and UF systems. As Figure 2 indicates, membranes can be produced in different physical forms.

The flat sheet can be employed directly in the plate & frame and spiral-wound configurations. Membrane has also been produced as a flat sheet and then slit and formed into small cylinders inside a suitably drilled or

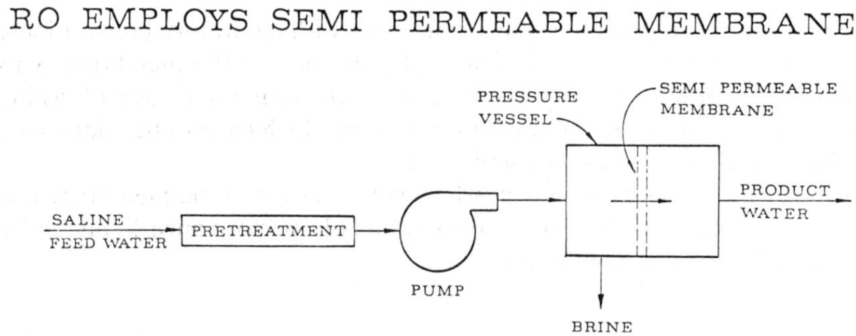

FIGURE 1

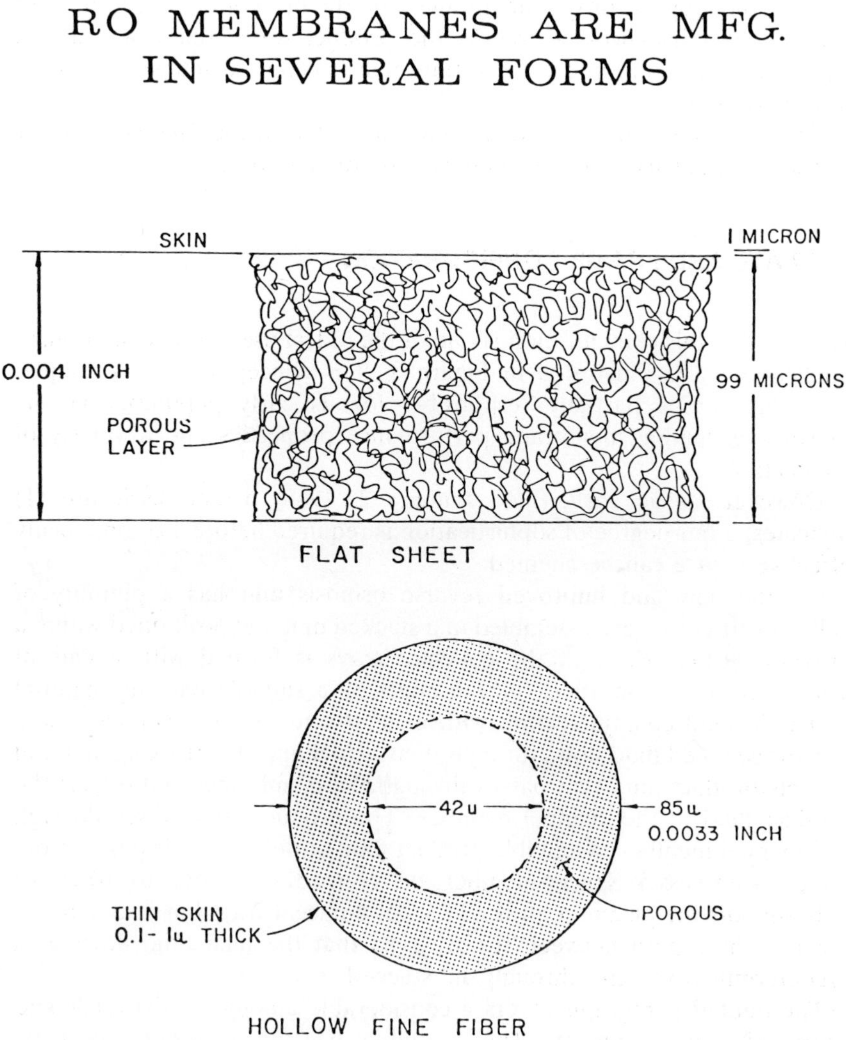

FIGURE 2

porous tube. Alternately, tubular membranes have been produced on a porous tubular support in-situ.

In any event, the spiral, tubular, and plate & frame membranes require external supports and product water removal channels in a suitable pressure housing.

The hollow fibers, while needing to be housed in a pressure tube, do not require support for the membranes and the bore of the fibers, in most (but not all) instances, serves as the product water removal channel, which is a limitation at the same time that it is an advantage, as will be discussed later.

These aspects will be discussed and illustrated in the following survey of the principal RO/UF membranes and configurations.

II. PLATE AND FRAME CONFIGURATION

Historically, this configuration is one of the oldest used in reverse osmosis applications. As illustrated in Figure 3, it is conceptually very simple, and that fact is partially responsible for its early popularity among those investigating membrane performance properties and a variety of applications.

However, as an early (filed January 2, 1968) patent disclosure (1) indicates, a fair degree of sophistication is required before a commercially effective device can be claimed:

"...the new and improved reverse osmosis unit has a plurality of collector trays that are assembled in a stacked unit and positioned within a pressure vessel. Each of the collector trays is formed with a pair of semipermeable membranes spaced apart by a suitable backing material that is in fluid communication with at least one channel member. As a pressurized feed fluid flows through the stacked unit, the backing material collects product fluid that passes through the membranes and directs the product fluid to the channel member. The product fluid passes through the channel member to suitable product fluid outlets where it passes from the pressure vessel. Spacer members between adjacent collector trays not only support the planar central regions of adjacent trays but also define a tortuous fluid path between the trays so that the feed fluid flows at a predetermined velocity through the stacked trays."

The quoted paragraph covers a considerable amount of materials and engineering development. The "suitable backing material" must be porous (low loss), thin, compression-resistant, etc., etc. The "spacer materials" geometry must provide the "tortuous path...at a predetermined velocity." Simply stated, this means that the velocities have to be high enough, in conjunction with the spacer geometry, to promote sufficient turbulence to minimize concentration polarization at the membrane surface without developing unacceptably high differential pressure losses (parasitic). Since the feed channel differential pressure

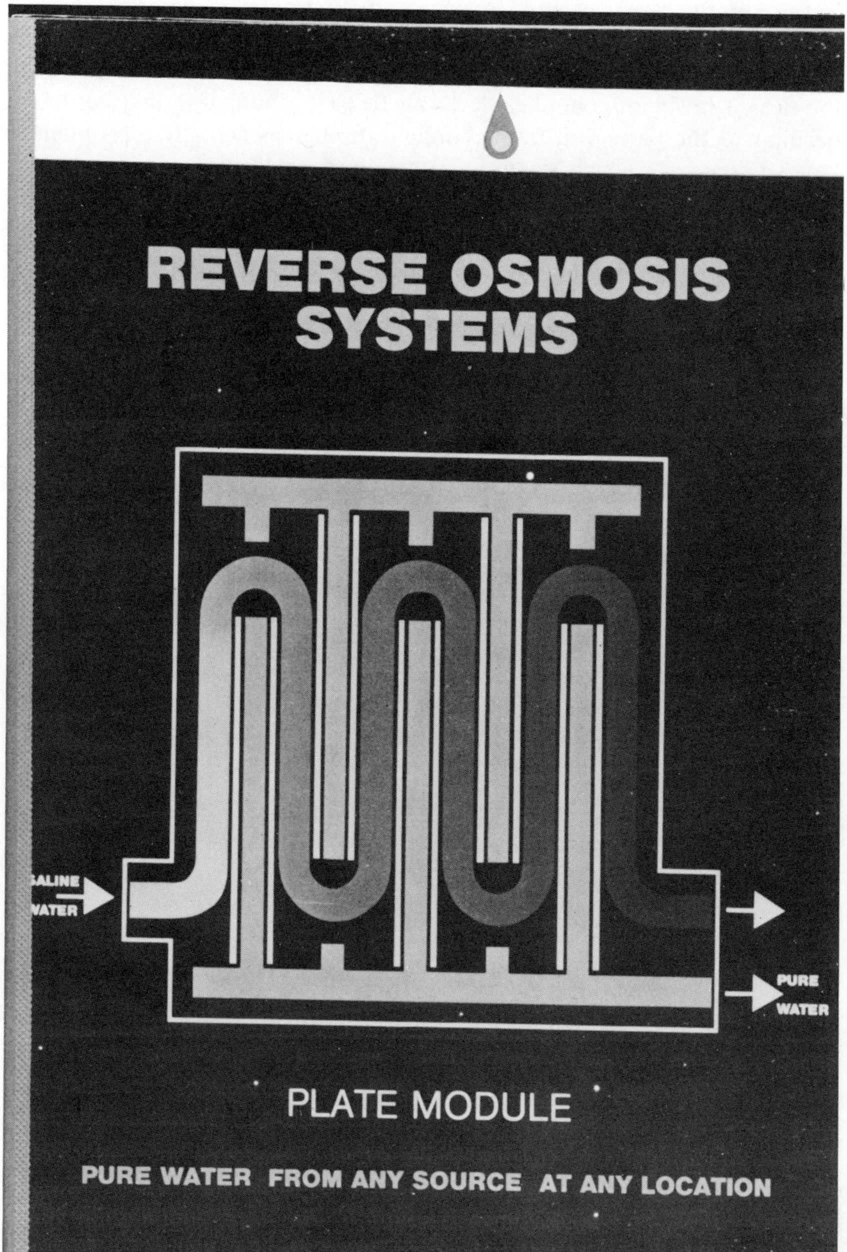
REVERSE OSMOSIS SYSTEMS
SALINE WATER
PURE WATER
PLATE MODULE
PURE WATER FROM ANY SOURCE AT ANY LOCATION

FIGURE 3

losses are the work of mixing necessary to minimize polarization, the need to minimize parasitic pressure drop losses and the need to minimize polarization pull the designer in two opposing directions. This aspect of the design problem is applicable to all RO/UF configurations...it is not peculiar to the plate and frame configuration. It is actually a problem of larger magnitude which severely limits some applications for the hollow fiber configuration.

Figure 4 illustrates a "module" (a plate and frame assembly enclosed in a sealed pressure tube), an internal flow schematic, and some details of the spacer, membrane, and support plate components and assembly. It should be enough to indicate that the fabrication of the components and the assembly of the module is quite labor-intensive and, accordingly, costly.

However, on the other hand, if one can stand the down time, and the relatively large capital costs per unit of throughput, the plate and frame configuration is perhaps the most suitable for food processing where perishability and the need to sanitize are paramount considerations. An early approach to this was described in a report entitled "Wurstack — a Reverse Osmosis Unit for Field Use" (2), published in 1967. Figure 5 shows the same equipment and detail as in Figure 4, plus details with

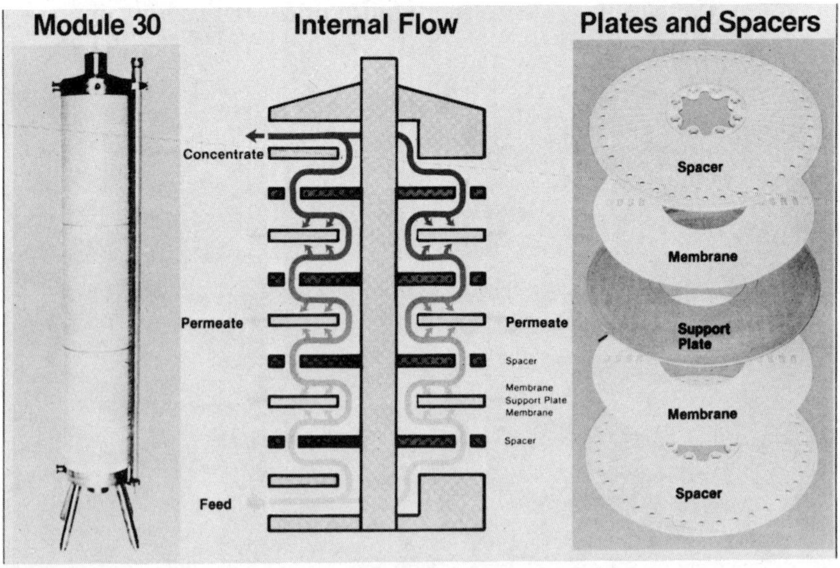

FIGURE 4

Hyperfiltration Modules and Membranes

DDS manufactures modules and membranes, as well as supplies complete plants ready for production, thus assuring the reliable performance of the entire system.

Modules

DDS manufactures a standard module, HF 30 with a diameter of 30 cm (11 13/16″) and 19 m² (205 ft.²) membrane area.
The module consists of spacers and membrane covered plates mounted in sandwich fashion and held together by a centre bolt. The vertical stack of spacers, plates and membranes is arranged so that membrane flow channels form under hydraulic compression. During operation the liquid flows over the membranes. The flow is divided into two streams, one called the permeate, passes through the membranes, and a stream which is retained by the membranes (known as the concentrate). The modules can operate at a pressure up to 8 MPa (1,160 psi) but usually operate at 4 to 5 MPa (580 to 725 psi).
The module is built of stainless steel and highly resistant polymer materials.

Membranes

The membranes used for hyperfiltration are usually made of cellulose acetate. This membrane type has proved itself to be reliable and economical. DDS manufactures a wide range of membranes of varying permabilities (defined as the percentage of dissolved solids passing through the membranes).

Sanitary Design

Since the design and construction materials of the module are stainless steel and polymer materials, it lends itself to quick and easy CIP cleaning. Whenever food is processed, a sanitary design of the plant is used.

FIGURE 5

respect to the operating pressures, materials of construction, and its "sanitary design" as it lends itself to "quick and easy...cleaning."

In summary, the plate and frame unit currently appears to be employed almost exclusively in relatively low flow capacity, high product value applications where its high costs and size requirements and limitations are of less consequence to its attributes for those applications.

III. TUBULAR CONFIGURATION

Some of the earliest work in reverse osmosis for water purification was carried out by Loeb, Sourirajan, and their associates, followed by field work on brackish well water at Coalinga, CA. This was based on work at the University of California at Los Angeles, which led to the development of asymmetric cellulose diacetate membranes (3).

The configuration employed at Coalinga was what is called tubular. Thus followed initial work with a plate and frame pilot unit. This is described by E.H. Sieveka (4) in part, as follows:

"To describe the unit briefly, it operates at 600 psi and treats a brackish feed water of about 2500 ppm TDS... The feed-brine rate is 10,800 gpd from which a nominal 5000 gpd of potable water is produced. The pressure vessels from this plant consist of 112 tubes, 10 feet long and 1 inch in diameter... The pressure tubes are conventional copper pipe... The cellulose acetate membrane is formed into a tube that fits inside the copper pipe. When the membrane tube is ready to be inserted, it is wrapped with two layers of nylon net material and one layer of filter paper so that when the assembly is complete the filter paper and nylon net will form a layer between the assembly is complete the filter paper and nylon net will form a layer between the membrane and the wall of the copper tube. The purpose of this arrangement is to provide a lateral flow path for the product water to the small holes placed 3 inches apart along the length of the tubes. In the pilot plant, as initially installed, loose-fitting plastic sleeves were wrapped around each copper tube to collect the product water. This was ideal for experimental purposes since each tube could be monitored separately. In a product plant a collection basin below the unit, or some other design, would doubtless be considered."

The feed flow to this unit on the basis of the above numbers would be approximately 7.5 gpm even if all of the tubes were connected in series, an unlikely and undesirable mode. The concentrate exit flow on the same basis would be approximately 4 gpm. These rates are so low in a 1" diameter pipe that they would almost certainly result in near-stagnant laminar flow with resulting severe polarization and possible scaling problems. The total membrane area available from the 112 tubes is only 291 sq. ft. as each tube contains 2.6 sq. ft. of active area, at most. As a result, this configuration requires, for most applications, undesirably high spatial requirements and other capital costs per unit of membrane area or permeate output.

The copper tubes were soon replaced by stainless steel or by various reinforced braided or filament-wound plastic approaches. The stainless steel approach, although expensive, could be expected to be acceptable or even demanded for dairy processing, particularly cheese whey. The 0.5″ internal diameter tube, 8 feet long, soon appeared to be the preferred configuration for rigid tubular supports. This provided 1 sq. ft. of membrane area per 8 foot length.

While this improved the velocity and polarization conditions, most tubular systems required considerable recirculation to boost velocities and recovery, resulting in differential pressure losses that were substantial, possible unwanted heating effects, and generally high pumping power and capital costs.

In an attempt to minimize the polarization problems with minimum pumping cost penalties, turbulence promoters were introduced which in effect diminished the effective internal diameter of the tubes, resulting in apparent greater superficial velocities for the same volumetric flow rates. These turbulence promoters took several forms. One of them is illustrated in Figure 6.

Even as velocities increased with recirculation, it was evident that on most of the feeds fouling was a major concern and there was almost always a significant need for some combination of pretreatment and cleaning technique. To combat this problem, some manufacturers proposed and offered a "sponge-ball" cleaning technique. It was claimed that it allowed the ball (whose diameter was close to that of the membrane-coated tube I.D.) through the system from beginning to end, presumably sweeping all foulants within its path eventually out the brine or concentrate exit port. There is little documentation with regard to the success of the various devices over a suitable range of feed problems. (See Fig. 7)

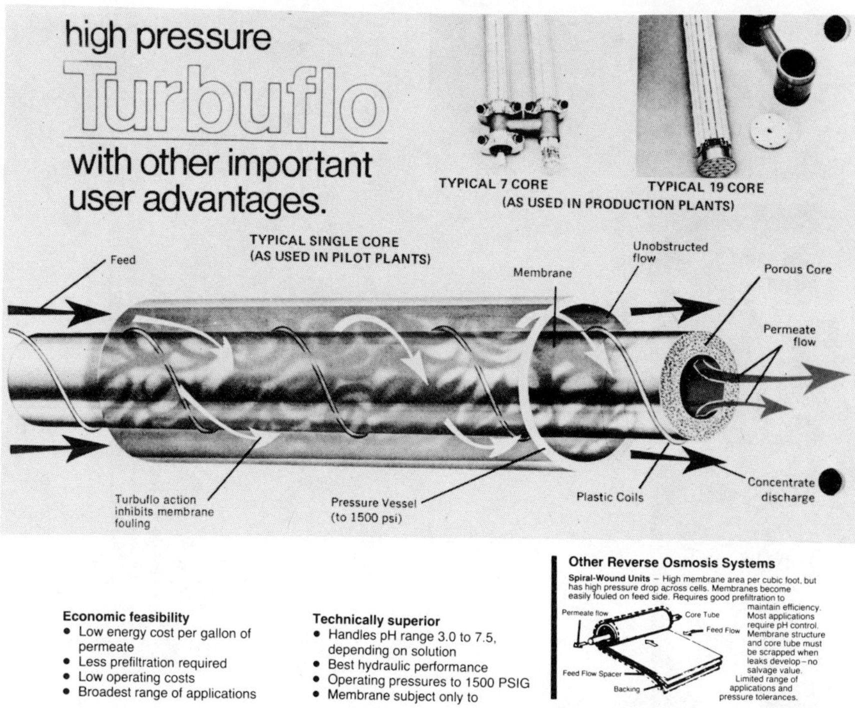

FIGURE 6

For low flow or high conversion applications, American-Standard Reverse Osmosis Systems include spherical turbulence promoters within the tubular elements. They create additional turbulence in the liquid flow and thereby minimize the degree to which polarized matter concentrates at the membrane surface, as well as reducing the possibility of membrane fouling due to the deposition of suspended matter.

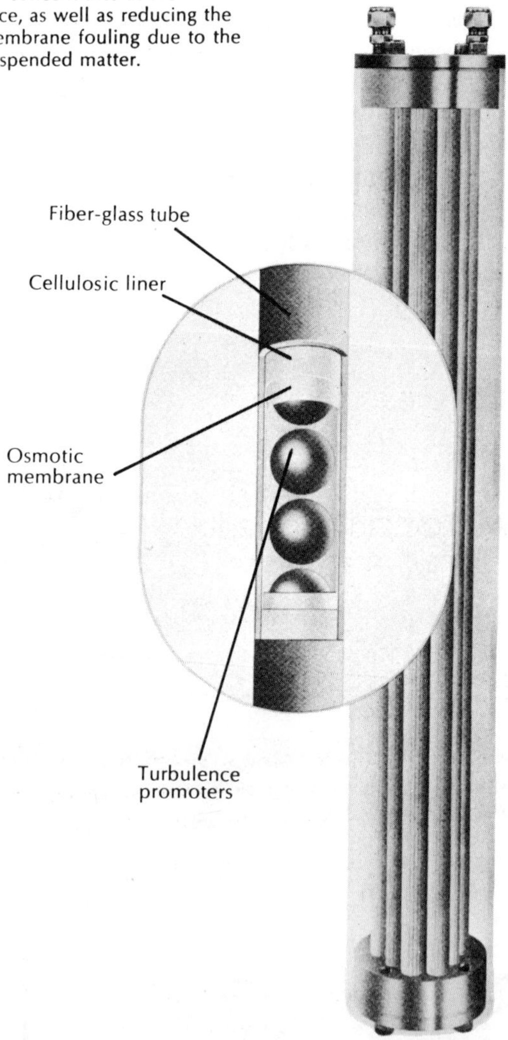

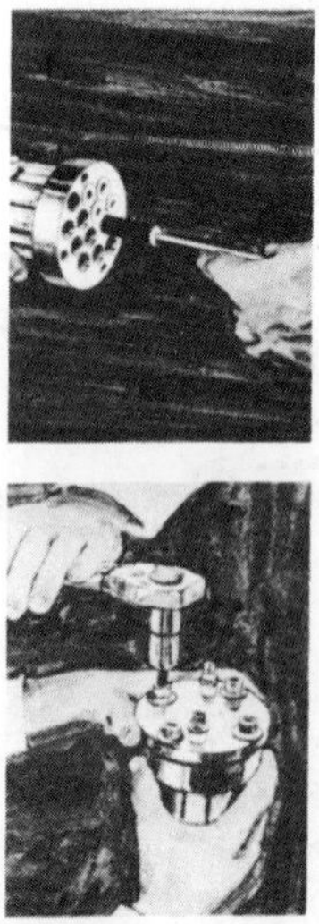

FIGURE 7

Several manufacturers alttempted to cut capital costs by introducing so-called mass production techniques to the membrane/element fabrication procedure which may be summarized as follow:

a) Manufacturing tubes by a filament-winding process similar to that used for fishing rods or golf clubs. In this case, the resin impregnation was controlled so that porosity remained sufficient to permit product water to pass through without excessive pressure drop. The membrane was coated on the interior surface of the tube in-situ.
b) Manufacturing tubes by a "sand-log" technique, where the structure was a resin-bonded sand sufficiently porous, having been cast around a core which, when removed, left a sufficiently smooth interior surface to allow casting of the membrane dope solution.
c) Manufacturing a continuous hose, with braided fibers also impregnated to allow sufficient residual porosity, followed by continuous casting in-situ on the inner surface...this ended up with two versions...the so-called "spaghetti" process and the "garden-hose process" (see Figures 8 & 9).

Figures 10 and 11 show a modern tubular system for ultrafiltration of wine. The tubes are the resin-bonded fishing rod version, while the piping and housings are stainless steel. Here, the choice of membranes is very important in terms of the retention characteristics required for this high value product, an application where relatively high power and unit capital costs are apparently quite tolerable. The largest units with a nominal rating of $10\,m^3$/hr. contain 378 tubes in 54 modules.

Tubular units have been employed in a few instances on a somewhat larger scale for processing of waste waters containing difficult suspended and colloidal solids, e.g., paper mill and potato-washing waste waters. At the present time it appears that the early hopes of the developers of the tubular products, that they could successfully acquire a relatively large volume of municipal, industrial, and waste water treatment at prices competitive with the spiral and hollow fiber configurations, have not materialized, and there does not appear to be any real prospect for a future tubular role in those areas. On the other hand, the specialty applications, such as wine processing, continue to be areas of apparent opportunity for tubular membrane equipment manufacturers who provide the applications and marketing effort required.

IV. HOLLOW FIBERS

It is of interest here to quote briefly from an early compendium (5) published in 1966 as "an effort on the part of one group of technicians to

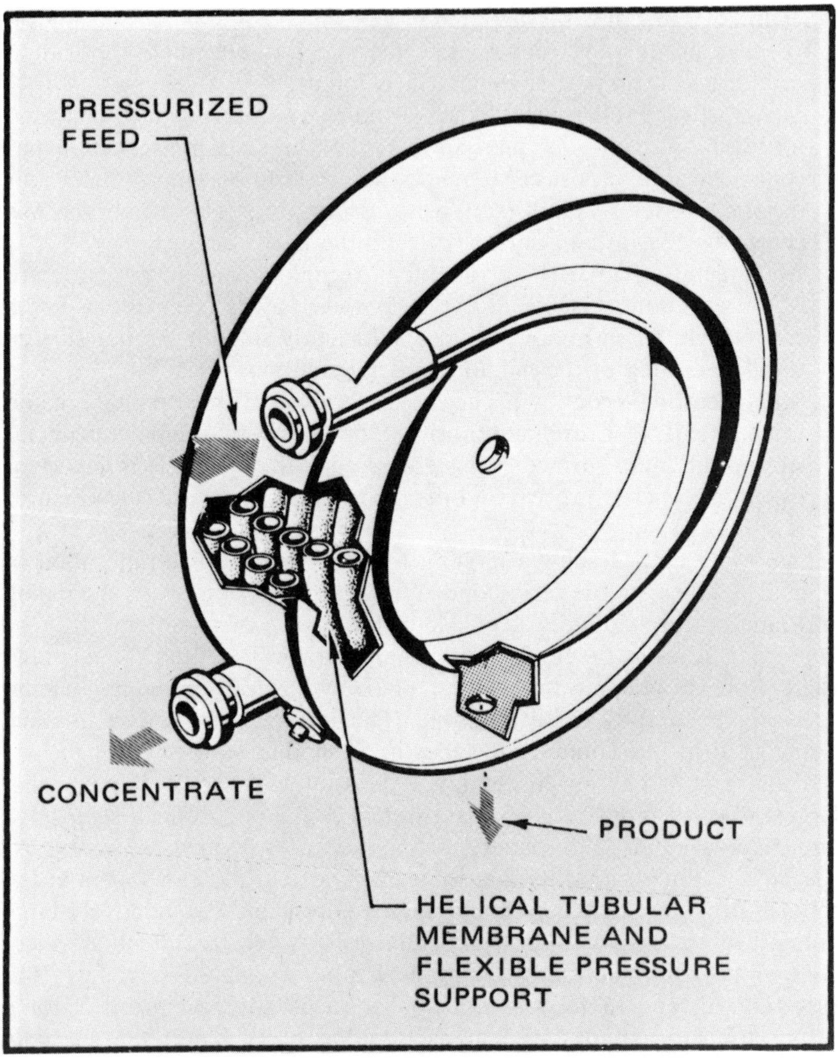

FIGURE 8

share what they have learned about reverse osmosis." In a Section entitled "Engineering of Reverse Osmosis Plants", Don Bray (one of the inventors and authors of a then contemporaneous basic spiral reverse osmosis element patent) briefly reviewed "Design Concepts" with respect to the features, advantages, and disadvantages of each of the configurations. The paragraph covering hollow fibers is headed "Small Tubes":

Philco-Ford RO System with Instrumented Console

FIGURE 9

"This approach utilizes small diameter tubes, where the walls of the tube serve both as the membrane material and the pressure containing unit...tube diameters are in the range of 5 to 100 microns with diameter-to-wall ratios of 4-to-5. Tubes are arranged in bundles with the ends sealed in headers by some material such as epoxy...because of the large surface area per unit plant space it may be possible to utilize this material in place of high flux membranes...the advantages of this approach are... the elimination of the need for a membrane support material, and the possibility of achieving high water fluxes before boundary-layer problems become limiting.

Its disadvantages are the need for good feed water filtration (probably a diatomaceous earth filtration)...and the need for short tube lengths for high flux membranes...it seems likely that the tubes...will be restricted to use at relatively low system pressures (500 psi)."

First, we want to remind the reader that the "tubes" referred to above have long since been referred to almost exclusively as hollow fibers although, as we shall see later, there may be a question as to how large a

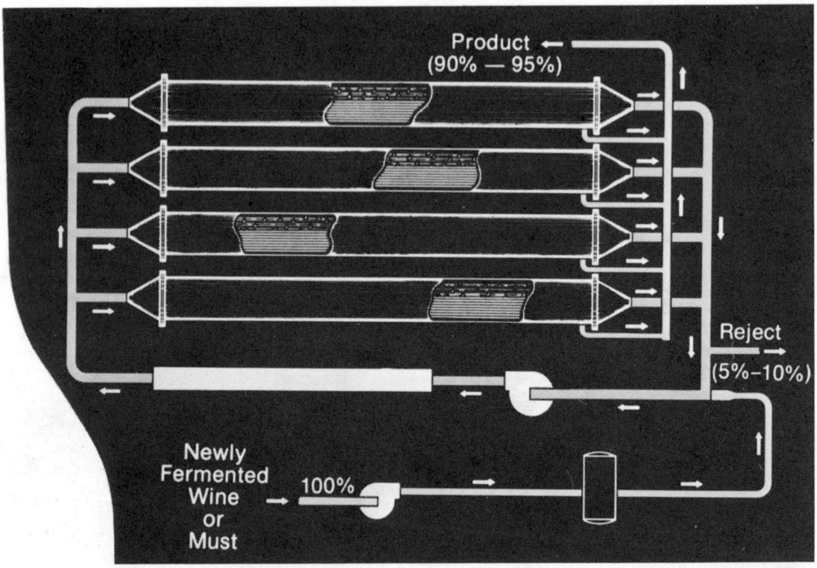

FIGURE 10

hollow fiber must be before it is classed as a tube or, conversely, how small must a tube be before it is classed as a hollow fiber? What else has happened since that statement was published?

At this writing, the hollow fiber configuration holds a substantial share of the reverse osmosis market, although largely limited to certain applications areas. There are two American manufacturers and one foreign manufacturer. The 500 psi upper pressure limitation foreseen by Bray has been overcome in that hollow fiber membrane elements are offered and utilzed for seawater conversion (at pressures up to 1000 psi).

The other caveat expressed by Bray, "the need for good feed water filtration", has had a significant limiting effect on the ability of the hollow fiber manufacturers to significantly penetrate or hold substantial applications areas of the reverse osmosis market. This is perhaps best exemplified by a large sector of the industrial water market, known in the trade as "Ultra-pure".

The nature of the always increasingly-stringent semi-conductor industry water requirements requires that a reverse osmosis membrane be employed at an early stage of the process. This requirement goes far beyond the generally modest demineralization requirement that applies

FIGURE 11

because the great majority of these plants operate on urban area municipal tap water (low TDS) feed sources. Those are supplies of surface origin, necessarily chlorinated, and they contain measurable amounts of humic and fulvic acids and other colloidal materials.

These materials cannot (at the offending particle size ranges) be significantly removed by granular media filtration. Taking cognizance of the earlier description of the hollow fiber configuration as one featuring extremely high packing density (and corresponding fine channel interstices), they are more susceptible to fouling and more difficult to clean. Putting it another way, if there is a partial blockage of the feed channels, it cannot be allowed to proceed too far before the cleaning flow in effect "channels" and cleaning is ineffective and fouling rapidly builds up.

The net result of this is that in the totality of RO market areas, the hollow fibers have a significant market share but it is generally limited to seawater and to brackish well waters where the fine colloid fouling problems are either absent or very much less. Those applications which operate on surface orgin feeds, waste waters, etc. are almost entirely handled by the spiral wound configuration. This market apportionment is recognized by insiders as well. . .the major hollow fiber manufacturer recently acquired a spiral capability.

It is beyond the scope of this paper to dwell at greater length on the strengths and weaknesses of one configuration vs another. But it is useful and necessary to understand to the extent practical the functional and engineering characteristics of the various configuration in terms of the available range of application conditions.

Figure 12 is a schematic illustration of a permeator which contains "several hundred thousand hollow fibers". While there are some differences among the manufacturers with respect to certain features, all of them introduce the pressurized feed water along a slotted or perforated central tube from which the feed flows radially through the fiber bundle to the periphery where it is discharged through a port as reject (concentrate). One must consider that for an 8″ diameter permeator operating at up to 75% recovery, the feed concentration can increase by a factor of 4 in a 4″ (the radius) path length. This is obviously a very low flow (laminar) condition. When this is coupled with the large mass of tightly packed fine fibers, the reason for the sensitivity to polarization and very fine particle (colloidal) fouling and the possible constraints on effective cleaning and performance restoration are obvious.

CONSTRUCTION

HOLLOSEP contains one or two elements in cylinderical vessel. And one element is composed of several hundred thousands hollow fibers.
Pressurized feed water is introduced into the flow path on the central axis, and it flows outside of hollow fibers in the radial direction.
The product water flows inside part of the hollow fiber to the open end, then, is collected and discharged from the vessel as effluent.
The concentrated water flows outside of the hollow fibers, and is collected and discharged from the vessel as brine.

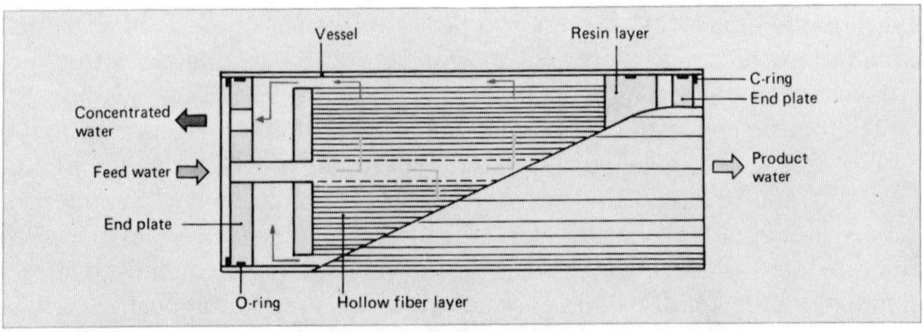

FIGURE 12

In this configuration, the product water path length must be kept as short as possible because of the high intrinsic resistance of the capillary bore which serves as the product water channel within the device. By some estimates, as much as one-third of the net driving pressure available may be used to counter this resistance which also manifests itself in an apparent lower bulk rejection compared to the intrinsic (higher) membrane rejection. This loss is minimized by appropriate choice of shell length combined with the "U-shaped" arrangement of the hollow fibers at one end so that the permeate water path length is only that of the shell length. When all is said and done, this aspect may be unknown to the end user who pays on some price-performance basis for the package. Figures 13 & 14 provide cutaway views in some detail of "CTA" and "Aramid" hollow fiber permeators.

As Table I illustrates, the hollow fiber permeator manufacturers have developed relatively large scale products in conjunction with appropriate marketing and applications support. This enables them to take continued cognizance of their product applications limitations while using their generally excellent membrane performance properties to maintain a strong competitive posture.

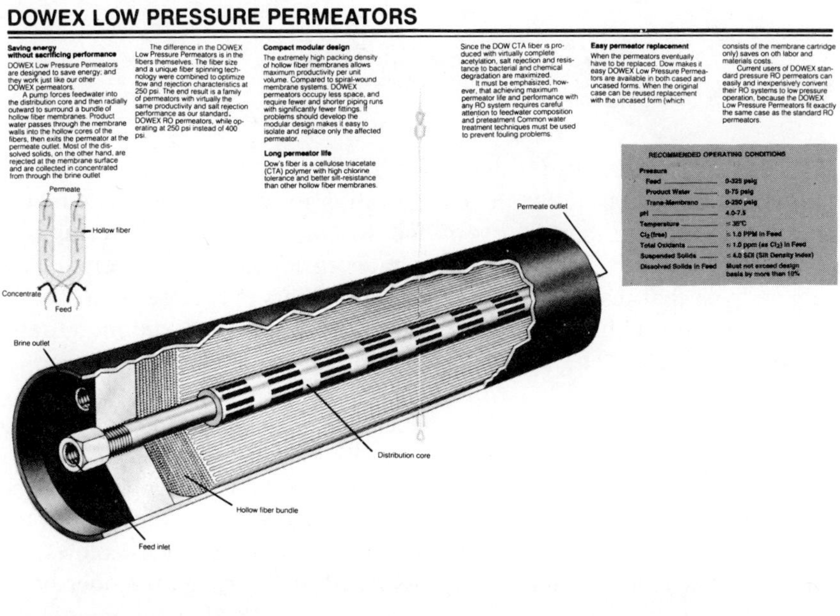

FIGURE 13

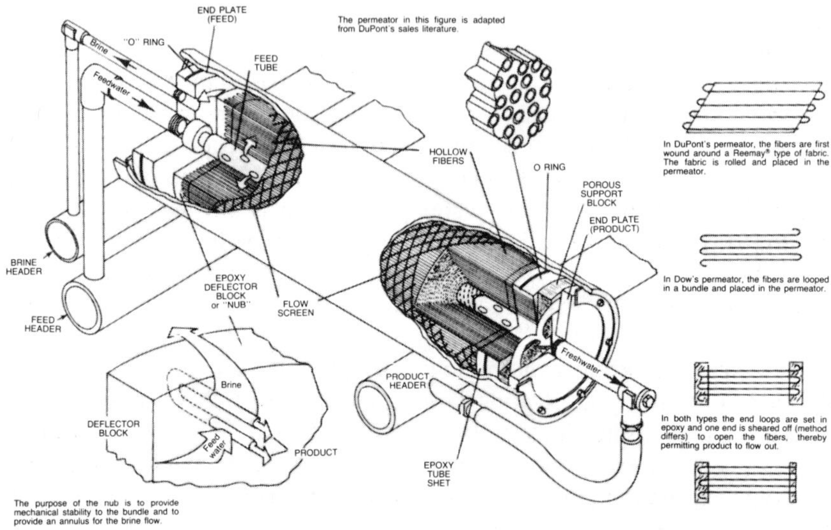

FIGURE 5-9. Permeator assembly for hollow fine fiber membranes.

FIGURE 14

V. LARGE HOLLOW FIBERS (SMALL TUBES?) FOR ULTRAFILTRATION

The reader can decide for himself which descriptive nomenclature he believes should be used for the configuration to be described. The manufacturer (Romicon, subsidiary of the Rohm & Haas Co.) terms it a hollow fiber configuration which operates in a somewhat different flow regime from that just described for the finer fibers in reverse osmosis use.

The "Romicon" fibers utilize an internally skinned polysulfone ultrafiltration membrane. The approximate fiber dimensions are as follows:

Bore Diameter — 1000 μ — Wall Thickness — 500 μ — O.D. — 2000 μ

These dimensions can be contrasted with those shown in Table 1 for the "Permasep" and "Dowex" reverse osmosis hollow fibers; it will be noted that they are bigger by a factor of approximately 20. This allows the different flow mode, running the pressurized feed through the fiber bore, at comparatively high velocities. The claim is made that the membrane

TABLE I

Reverse osmosis hollow fiber "fact sheet"

Mfr.	Product	Fiber dimensions (μ)	Shell & case dim. (in.)	Permeate output rating (GPD)	Recovery (%)	Membrane	Max. limit HOCl, HOBr
DuPont	B-9 (Brackish)	O.D. 85 I.D. 42	O.D. 10 Length 53	16,000–25,000	50–75	Asymmetric Polyamide ("Aramid")	0.0
DuPont	B-10 (Seawater)	O.D. 93 I.D. 42 Wall 25	O.D. 8 Length 59	6,300	30–35	"	0.0
Dow	Dowex 20 (Brackish)	O.D. 250 I.D. 90 Wall 80	O.D. 10.75 Length 48	20,000	50–75	Cellulose Triacetate	1.0

surfaces will thereby be less susceptible to fouling and that they will clean more readily. These features are especially important in many UF applications where the concentrated or dewatered product is the object of the process.

The cartridge and systems are illustrated in Figures 15, 16 and 17. Reverse flow of permeate is also possible which allows backflushing as a regular technique, as illustrated in Figures 18 & 19. This claim is made that the backflushing provides a positive means of freeing the foulant from the membrane surface, thus increasing cleaning efficiency and extending run times and cartridge life. As with the large tubes described earlier, this configuration is somewhat more expensive in capital cost than the small hollow fiber RO systems but finds its technical and economic niche when the value of the process and product are taken into consideration, generally in systems which have small capacity compared to RO systems using the smaller fibers.

VI. SPIRAL WOUND MEMBRANE ELEMENTS (RO & UF)

This concept was originally developed and reduced to practice starting in the early 1960's. The concept is cumbersome to explain in print, even with drawings. Even a rough model is more satisfactory. However, in this discussion, we refer to the chapter by Kremen (6) published in 1977 for such enlightenment as it may provide, as augmented in this text by Figures 20, 21, and 22.

"The components of the module are flat sheet or netting materials which are rolled around a central tube which serves both as a mandrel for the rolling operation and as the permeate water collection conduit...at the time of attachment, prior to roll-up, the feed spacer netting is between two layers of membrane so oriented that the barrier surface of the membrane is facing the netting. The feed flow enters the rolled-up module...under a suitable hydrostatic pressure. Permeate...is then forced through the membrane, through the membrane support sheet, and into the product water channel material. The feed water continues through the netting diminishing in volume and increasing in concentration through subsequent modules and vessels, until finally let out through a concentrate valving arrangement downstream in the system...

The permeate then flows through the product water channel material until it reaches the product water tube...perforated with holes of proper diameter and spacing to allow passage of permeate from the channel material into the product water tube, where it then goes to final collection and distribution.

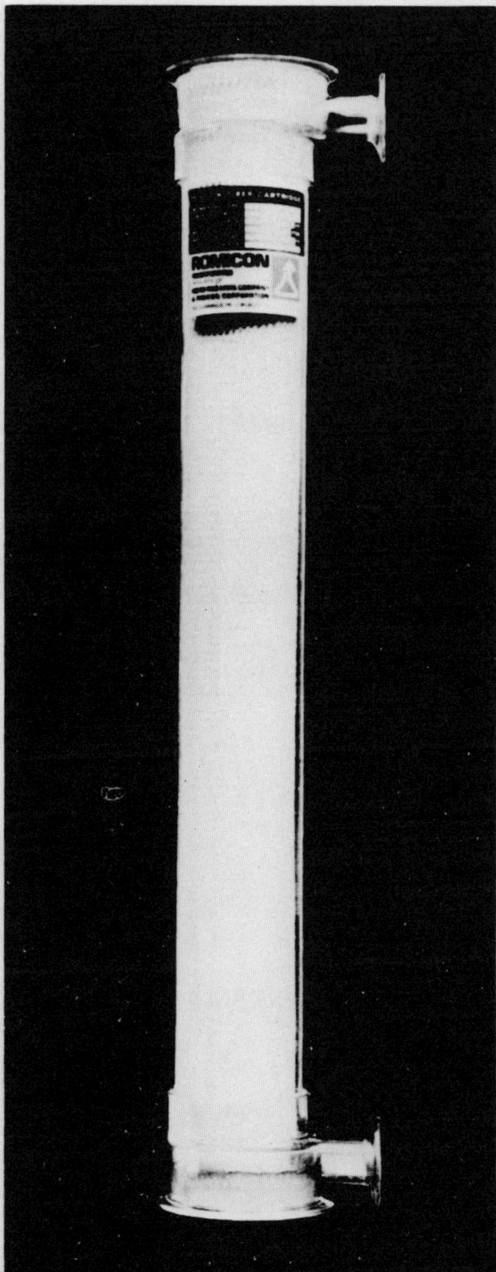

Hollow Fiber Cartridge

FIGURE 15

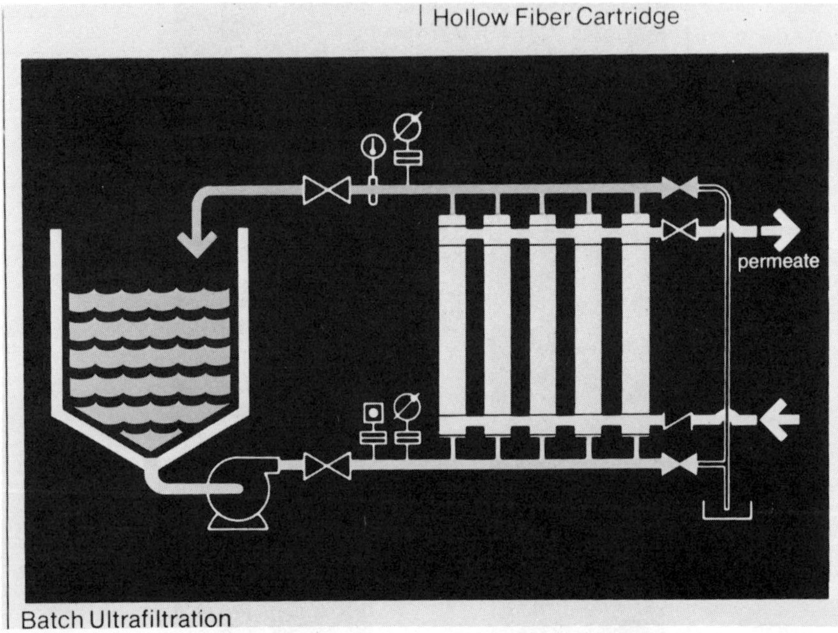

FIGURE 16

FIGURE 17

concentrate
permeate
permeate
spongy support structure
thin rejecting skin
fiber lumen
feed
Romicon Process Mode
High molecular weight solute and colloidal solids (rejected by the membrane)
Low molecular weight solute (not rejected by membrane)
Solvent

FIGURE 18

FIGURE 19

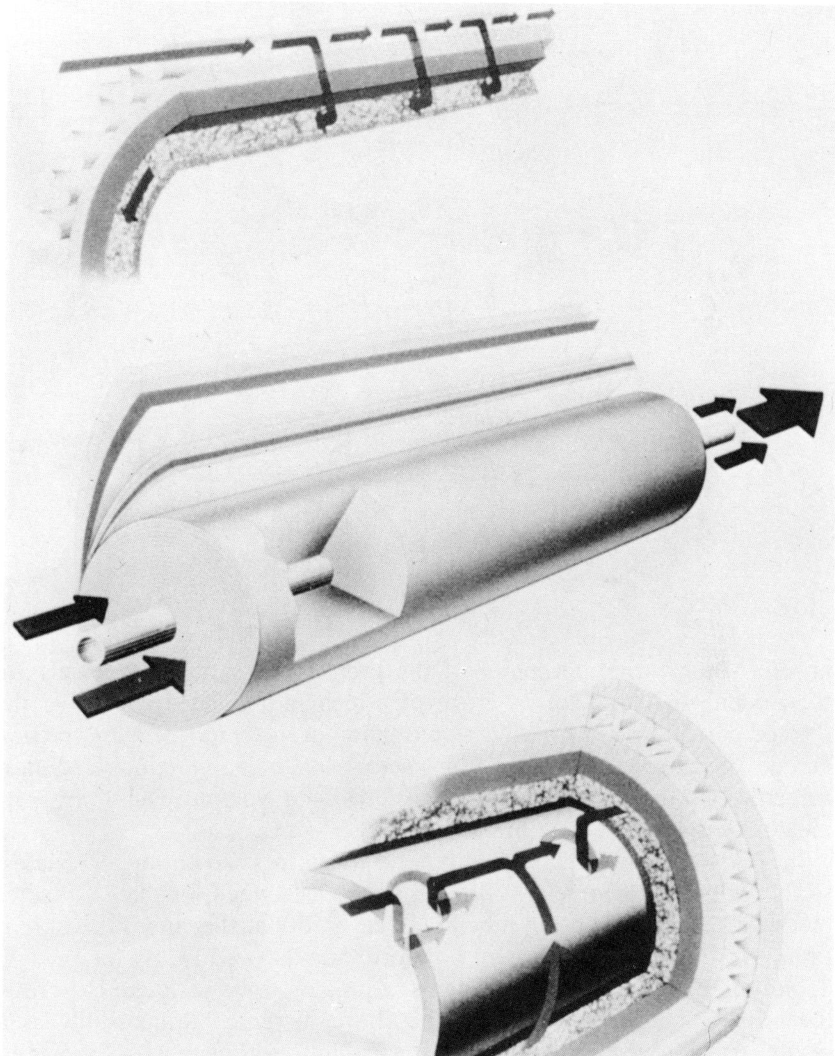

FIGURE 20

The rolled up module then resembles a plug-in cartridge filter element, complete with retaining outer wrap, and seals to isolate and direct the feed and permeate flows, when connected to other modules and inserted in a vessel...the modules contain circumferential chevron or U-cup seals which force the feed flow through the modules rather than allow it to flow

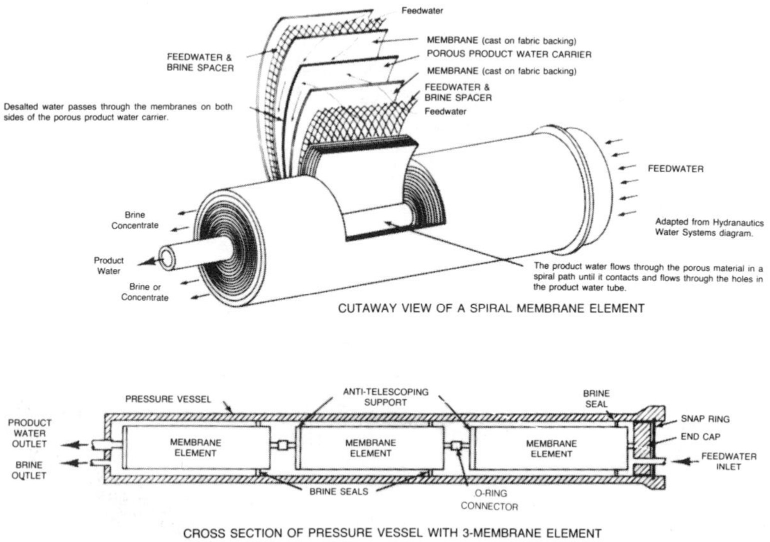

Spiral membrane-cut-away view with elements in a pressure vessel.

FIGURE 21

around the external perimeter of the pressure vessels...the vessels are then connected together in an array which is designed to achieve the desired permeate water purification fraction of the feed (recovery fraction and/or the desired concentration factor) while maintaining feed flow velocities in the face of the diminishing feed volume caused by permeation through the membrane."

In 1976, when Kremen's chapter (6) was in preparation, the latest spiral wound commercial offerings were described as "8" diameter modules..." Based on the membrane capability at that time, the output ratings for those elements were no more than 5000 GPD. By the present definition of spiral module (elements in pressure tube with end closures), the outputs and the rejection values have increased dramatically as a result of improved membrane flux capability with equal or improved rejection, plus fabrication techniques which have further minimized leakage and inefficiencies.

This is illustrate in Table II. This has laid the basis for low pressure membranes for brackish water purification and waste water reclamation, thus lowering product water costs, particularly as a result of the power cost savings thereby accruing.

The spirals, in addition to their dominance in the ultrapure water market, have been almost unchallenged in the area of waste water

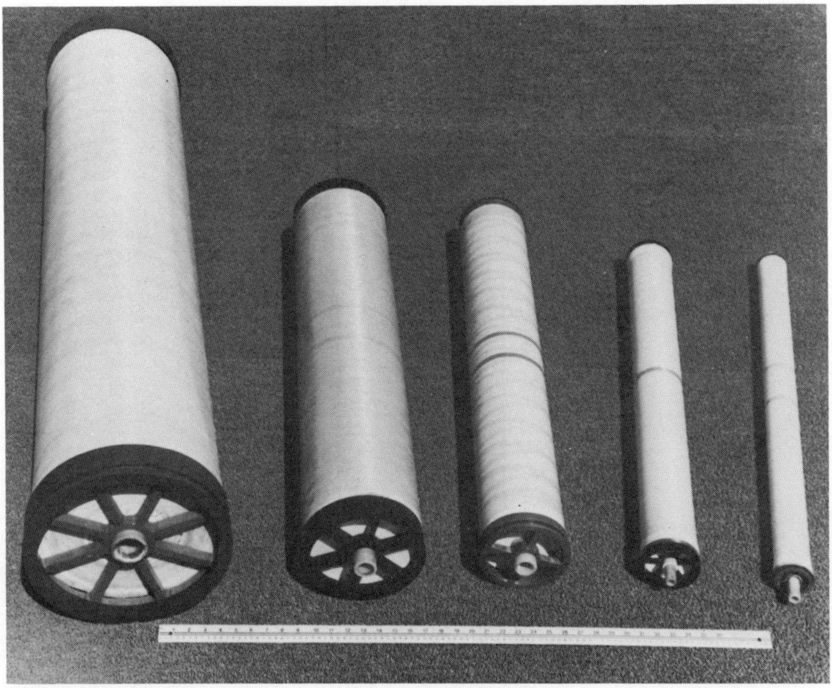

FIGURE 22

reclamation. This has been due to many factors which include, most importantly, those listed by Kremen (Ref. 6, p. 374):

"e. Very low labor cost and little downtime for membrane replacement in existing operating systems at the site.

f. Ability to combine modules and vessels in arrays which can adequately cover any recovery or concentration range required, to the extent that the feed chemistry permits, with proper provision for pressure drop losses, osmotic pressure increases, etc.

g. Ability to operate at reasonable levels of pretreatment with concurrent ability to be effectively cleaned at acceptable intervals.

h. Ability to utilize any membrane cast in sheet or film form, asymmetric or thin film composite."

Although the foregoing attributes were listed in 1976, they were almost precisely applicable to a 1982 situation which is described in a paper by David Paul (7) in May 1984. The large complex of power plants at the San Juan Power Station of Public Service of New Mexico generated large volumes of waste water in a watershort region where it was essential to

TABLE II

Hydranautics reverse osmosis spiral wound element "fact sheet"

Model[1]	Membrane	Permeate output rating (GDP)	Per cent Salt rej.	Max. limit HOCl, HOBr	"Module"[2] rating
CAB 8540A	Cellulose acetate blend (asymmetric)	9,500	95	1.0	50,000 GPD @ 50% Recovery
CAB 8540B	″	7,600	98	1.0	40,000 GPD @ 50% Recovery
IFC 8540	Polyamide type thin film composite	11,800	98.5	0.0	60,000 GPD @ 50% Recovery

[1] All elements are 40″ long; O.D. = 8.5 in.
[2] "Module" consists of 6 elements assembled (with end closures) in a standard FRP pressure tube whose dimensions are: Length = 22 ft. I.D. = 8.5 in.

protect the water table. This led to the imposition of a zero discharge requirement wherein no waste could exit the plant boundaries, and it was determined that for the 1500 gpm of waste generated no more than 30 gpm could be handled by return to the mines or solar ponding.

The treatment train for the widely varying waste stream composition and volume is shown in Figure 23. It was installed in 1980. It included stabilization and equalization, temperature control (lower limit), lime-soda clarification for partial silica and hardness removal, granular media and micron filtration (protective), reverse osmosis to reclaim 1,200 gpm of good quality water for re-use, and a seed slurry vapor compression evaporator (brine concentrator).

The pretreatment and reverse osmosis system were installed in 1980. It utilized composite non-cellulosic spiral membrane elements in an array designed to permit handling possible high TDS waste water feeds at the 80% permeate recovery level (five-fold concentration).

It was soon obvious that major problems had developed from the presence and growth of algae in the holding pond and throughout the system. This led in part to the decision to replace the initial element loading with cellulosic blend membrane elements, so that the feed water could be continuously and effectively chlorinated wherever required throughout the system. The replacement elements, reflecting continuing improved performance capability, had significantly higher flux ratings

FLOW SCHEMATIC - SAN JUAN POWER STATION

PUBLIC SERVICE OF NEW MEXICO WASTEWATER TREATMENT AND RECOVERY SYSTEM

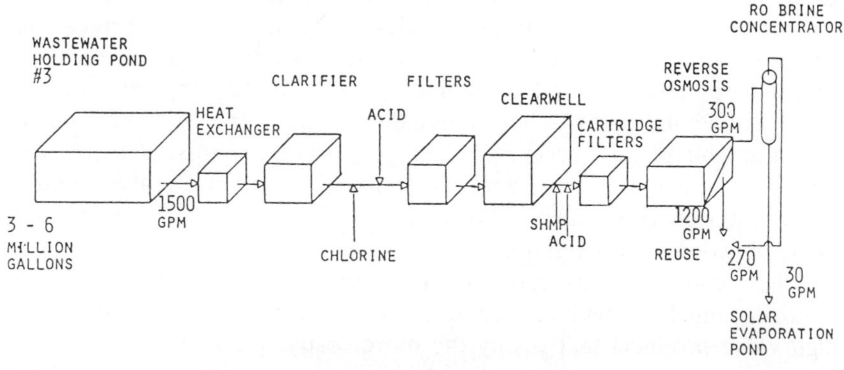

FIGURE 23

than the original elements so that lower pressures sufficed to produce rated flow at the then prevailing chemistry in a simplified array at the same overall recovery.

When it was decided to re-membrane the plant (there were four trains each containing 144 membrane elements in 24 pressure tubes), the required downtime for each train for membrane replacement was less than three days. This included considerable cleanup and refurbishing.

A modest technical support program monitored the status of the plant following reloading and produced recommendations, among others, for bank rotation and cleaning which have been highly effective, so that there has been little change in performance for almost 20 months at this writing. This ability to step into a previous grievously-troubled situation and produce a complete turnaround is entirely consistent with the spiral element virtues listed by Kremen (6) as quoted in a preceding paragraph.

To quote David Paul (7), "the first reverse osmosis system failed, the second is a great success."

The spiral market is served by at least six domestic manufacturers and one foreign manufacturer. As with other configurations, variations have been proposed (see Figures 24 & 25) but, in essence, the basic design, in much more efficient and massive form, persists, augmented by equally large improvements in membrane/systems capability (see Fig. 26). Those facts, combined with the advantageous engineering and applications attributes, have combined to keep spiral module pricing and costs at a very reasonable level.

VIII. SUMMARY

Over the past twenty years a number of configurations and membrane combinations have been proposed, developed to varying degrees, and accepted for a wide variety of reverse osmosis and ultrafiltration applications. For technical and economic reasons, the major configurations, spiral wound and hollow fiber, dominate the large plant business covering seawater conversion, brackish water purification, industrial water production to meet stringent standards, military mobile and seaborne units, waste water concentration and reclamation for re-use ranging from power plant blowdowns to irrigation return drainage, etc.

Other configurations have their niche in specialized applications requiring much individual technical support and generally dealing with high value products that justify the more costly systems.

Progress continues, practically across the board with regard to mem-

Basic Configuration

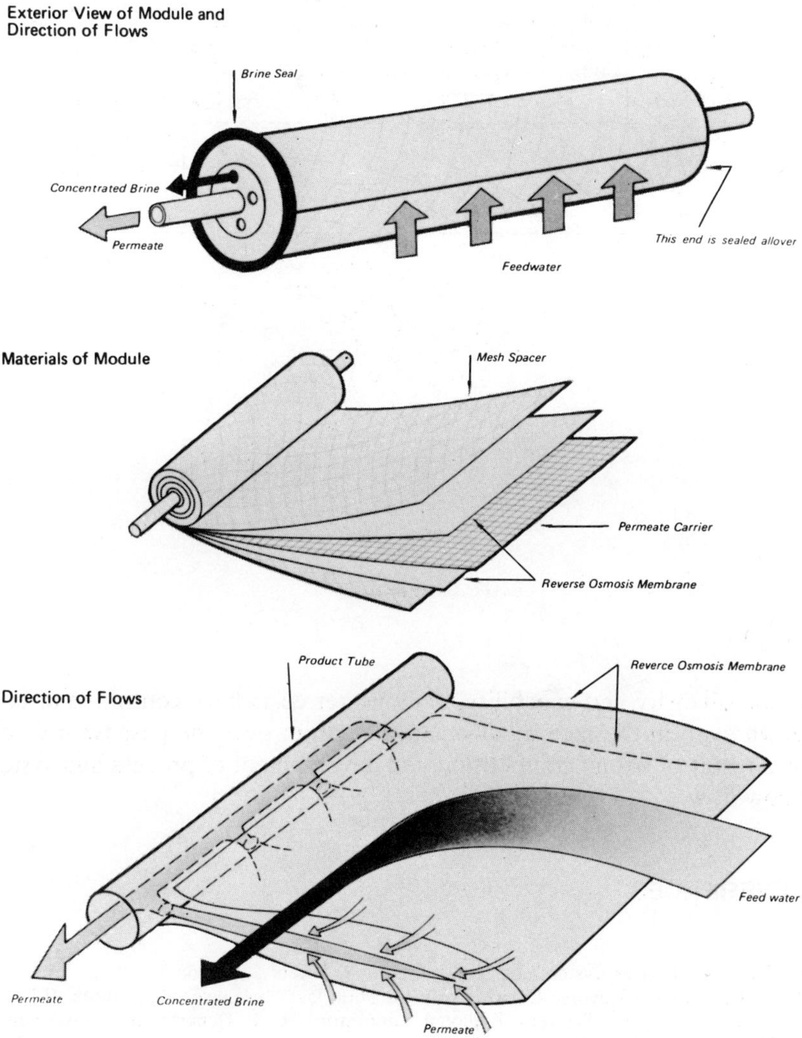

FIGURE 24.

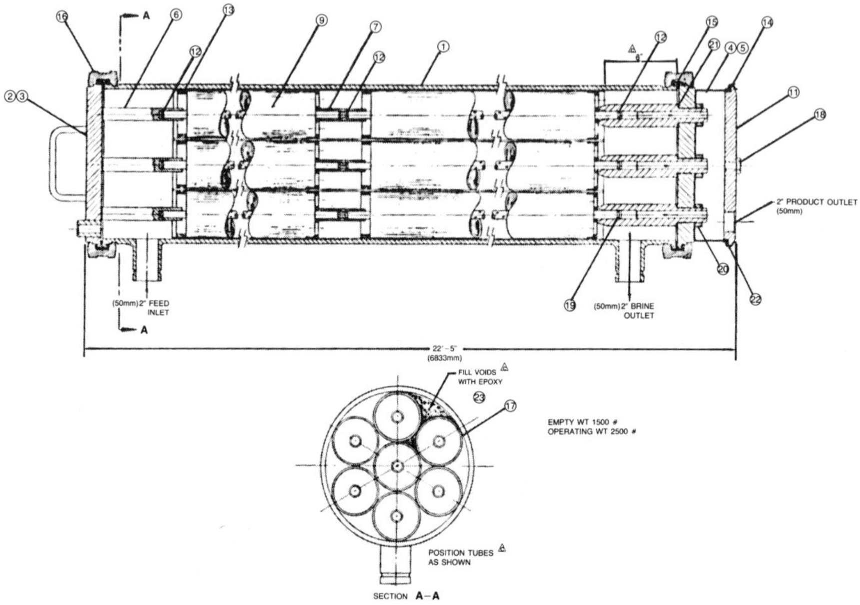

Figure 1

Single Cluster Assembly Drawing

FIGURE 25

brane diversity and capability, while water costs have continued to come down even in the face of substantial inflation over the past twenty years as a result of strong competition and development of process and systems know-how.

REFERENCES

1. "Reverse Osmosis System" by Kohl, A.L. U.S. Patent 3,494,468
2. "Wurstack — A Reverse Osmosis Unit for Food Use" by Lowe, E., Durkee, E.L., and Morgan, A.I., Jr. Western Regional Laboratory, U.S. Department of Agriculture, Albany, CA 94710, August 16, 1967
3. "Preparation and Performance of High Flux Cellulose Acetate Desalination Membranes" by Loeb, S. Chapter 3, "Desalination by Reverse Osmosis", Merten, U. (Ed.), M.I.T. Press, 1966
4. "Reverse Osmosis Pilot Plants" by Sieveka, E.H. Chapter 7, p. 245–6, "Desalination by Reverse Osmosis", Merten, U. (Ed.), M.I.T. Press, 1966
5. "Engineering of Reverse Osmosis Plants", by Bray, D. Chapter 6, p. 212–3, "Desalination by Reverse Osmosis", Merten, U. (Ed.), M.I.T. Press, 1966

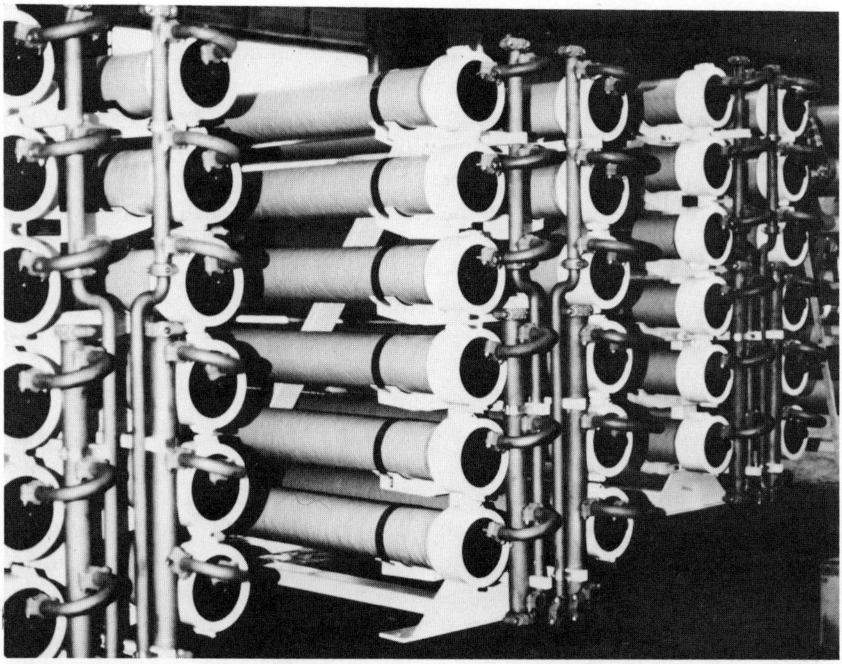

FIGURE 26 Hydranautics "Hydrablocks" spiral membrane and pressure tube rack assemblies.

6. "Technology and Engineering of ROGA Spiral-Wound Reverse Osmosis Membrane Modules" by Kremen, S., Chapter 17, "Reverse Osmosis and Synthetic Membranes", Sourirajan, S. (Ed.), National Research Council of Canada, 1977
7. "The History of Industrial Water Recycling to Achieve Zero Liquid Discharge" by Paul, D.H., Public Service Company of New Mexico, Volume 1, Technical Proceedings, Annual Conference of Water Supply Improvement Association, Orlando, Florida, May 13–18, 1984

DISCUSSION

QUESTION What's been the experience at Yuma with fouling problems that were serious?

ANSWER The answer to that is they had fouling problems that I don't think were serious. I think I'll use the blackboard. One of the problems at Yuma I think can be explained by this concept. I have a former manager of engineering who used to work for me at Fluid Systems who used to say that a slide rule is better than a calculator and that also applies to a computer because it doesn't require any batteries and it only yields three significant figures. When you have something that is overcomputerized and overstudied and you're going out to the third decimal place, you're really going to start being impractically critical. By the way, I can say this without too much fear of contradiction because the government hired me as a consultant at the time they had the Yuma proof-tester problem — if you want to call it a problem. Drawing: This is time and this, so called specific productivity. The manufacturers told the government, "Hey, we expected the climb line to look something like this and this might be four or five years out here. We expect, however, the drop dead tape to get you the water you want to be done here." And after running the proof test for about 6 months they ended up here. Now, if that happened in a commercial setting that would have been the end of it, everybody would have gone home happy and the customer would have had his water and the manufacturer would have gone about his business. The government, said, "We don't really know when we extrapolate this if it's going to be over here or whether it will continue like this." I think the government basically is responsible for some of this "problem" anyway in the sense they shot themselves in their own foot with their pretreatment procedures. You've heard of inline flocculation, well now you've learned about inline dispersion. They were redissolving stuff that they had previously coagulated. It was resolved basically by the manufacturers agreeing to provide about 15–20% more membrane area at no cost to the government. And that basically, on a normalized basis took this drop dead line down here. But, I don't think this was a serious fouling problem in a commercial sense. I really think it was partly self-inflicted because they went into a vast project, hundreds of millions of dollars, and they used a second-hand, cobbled-up pile of a clarifier. Have you ever seen a clarifier having stuff coming over the weir that has JTU's over 300? If you haven't, you should have been at Yuma, you would have seen it everyday, and that's not much of a clarifier.

ANSWER (Question inaudible) I guess I am glad you asked that question. The original patent Don Brey and Wes Moland planned expires next year. Does this make any difference? Is there any change in strategy, so to speak? Well, for one thing, the patent doesn't really bother anyone manufacturing in the U.S. The U.S. government had a right to license and sub-license with no royalty to anyone manufacturing in the U.S. And that's what's happening. And it's the Brey patent that's used; Moland is a single use patent. The patents overseas are all expiring in the next three or four years and some of them are already expired. To answer your question, "Will it have an effect on the business climate and will some overseas people rush into membrane manufacturing because there is no patent problem?" Well, I don't know. I guess the real question is would you want your son to become a reverse osmosis entrepreneur or would you invest your money in something else?

QUESTION Could you give us some idea just how big the reverse osmosis field is? How many square feet or gallons per square foot per day capacity are installed and at about what cost each year?

ANSWER I really can't give you a good number on that. I really lost track of that lately. I'm sure that the capacity that's out there is well in excess of 100 million gallons of daily production. I can tell you in terms of cost, a couple of ball park ranges, and we're talking large plants now. Whether or not the price will ever permit a profit, I don't know, but people are talking about installing a brackish water plant in the upper end of the well stream in mineral free treatment. Now, you're seeing numbers that I guess would be around $.60 an installed daily gallon. Pretty power cost sensitive of course. We're probably talking water costs which would be down below $.75 a kilogallon. Now on surface waters it's going to be somewhat more and its going to vary tremendously with the pre-treatment. On sea water, I think we're seeing numbers there that are in the range from about three to five dollars a daily gallon, depending on the sea water and the size of the plant, and we're seeing water costs at about three to five dollars a kilogallon, again depending upon the power rates. Sea water capability out there: it's hard for me to say, but I would imagine it's already up to 30 to 40 million gallons. Maybe Bob Peterson can have some better numbers.

The Role of Synthetic and Biological Membranes in the Transdermal Delivery of Drugs

DAVID J. ENSCORE AND JANE E. SHAW

ALZA Corporation, Palo Alto, California

The characterization of transport parameters of both synthetic and biological membranes is of critical importance in the design of transdermal drug delivery systems. The inherent permeability of a drug across human skin sets an upper limit on the dose of the drug which can be administered transdermally, within the flexibility of variation of system size. Moreover, the diffusivity and the apparent solubility of a drug in skin affect the time required for a delivery system to reach a steady state rate of drug input.

The choice of the synthetic polymer membrane for control of the rate of drug release from a transdermal system influences not only the rate of drug input, but also the variability, within and between patients, in the dose of drug administered. The contact adhesive layer of a transdermal system functions both as a diffusion membrane and as the means of maintaining contact between the system and the skin. Therefore, selection of a contact adhesive requires optimization between the adhesive and diffusive properties of this membrane layer.

The interaction among these membranes, which ultimately determines the rate of drug input to the systemic circulation, can be modeled using a relatively simple relationships drawn from the study of the transport of low molecular weight penetrants in glassy and rubbery polymers. These effects are illustrated using data attained for recently developed transdermal drug delivery systems.

INTRODUCTION

Over the last decade, the pharmaceutical industry has placed increasing emphasis on the development of novel drug delivery systems. A new class of dosage forms, transdermal therapeutic systems, which deliver drugs to the systemic circulation through intact human skin is one product of these extensive research and product development activities. Transdermal drug delivery for systemic therapy is now widely recognized as feasible for a variety of compounds. Moreover, several transdermal systems are either commercially available or nearing regulatory approval.

By using intact skin as the drug entry portal, transdermal therapeutic systems can deliver certain drugs to the systemic circulation in a manner superior to oral administration. Specific advantages of transdermal therapy include:

1. Elimination of the vagaries of gastrointestinal absorption;
2. Reduction of drug metabolism due to an initial bypass of the liver;
3. Safe administration of drugs with small therapeutic indices;
4. Elimination of pulsed drug administration with a corresponding reduction of side effects;
5. Utilization of drugs with short biological half-lives (1).

The transport of drugs, as well as most other substances, across skin occurs by a solution/diffusion process as described by Fick's Law (2). Additionally, drug release from all current transdermal dosage forms is by a diffusion mechanism. In the design of transdermal drug delivery systems, it is important to be able to characterize the transport properties of both the polymers selected for inclusion in the system and the skin itself. Standard experimental techniques and methods of analysis, widely used in the study of the transport of low molecular weight species in polymers, can be utilized for the study of drug transport in synthetic membranes and skin.

While there are several designs applicable to development of transdermal dosage forms, this paper will be limited to a discussion of systems containing non-porous drug release rate controlling membranes. A schematic representation of this design is shown in Figure 1. This design is composed of several elements; an impermeable backing, a fluid or polymeric reservoir containing sufficient drug for delivery over the system duration, the rate controlling membrane, and a contact adhesive layer for fixing the system to the skin. The contact adhesive is covered by a removable protective strip. An example of this design is the Transderm-Nitro® system jointly developed by ALZA and Ciba-Geigy.

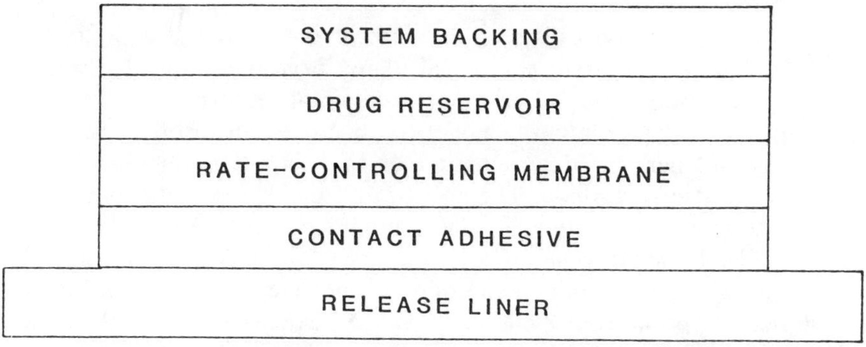

FIGURE 1 Schematic representation of a rate-controlling membrane transdermal therapeutic system.

MEMBRANE CHARACTERIZATION

The parameter that best characterizes the transport of a diffusing species in a polymeric membrane is the permeability coefficient of the membrane to the diffusant ($\bar{P}$). For a unit activity penetrant source, the permeability can be defined as the product of a kinetic parameter, the diffusion coefficient (D), and a thermodynamic parameter, the equilibrium solubility of the species in the polymer (C_S).

$$\bar{P} \equiv DC_S$$

The steady state flux of a penetrant across a polymer membrane (J) is related to the permeability through the membrane thickness (ℓ).

$$J = \frac{\bar{P}}{\ell} = \frac{DC_S}{\ell}$$

The steady state flux can be measured by imposing the membrane between a constant activity source of penetrant and a receptor operating at perfect sink conditions. Data from a typical permeation experiment are presented in Figure 2. They steady state portion of the permeation is preceeded by a transient period during which the steady state concentration gradient across the membrane is being established. The time axis intercept of the steady state portion of the permeation; termed the time lag (τ), allows calculation of the diffusion coefficient using:

$$\tau = \frac{\ell^2}{6D}$$

If ideal, Henry's Law dissolution behavior is assumed, the saturation concentration of penetrant can be calculated from the permeability (3).

For glassy polymers and polymers containing absorptive fillers, the assumption of ideal solution behavior frequently is not justified. In these cases, an independent determination of the nature of the dissolution process must be performed. The polymer is equilibrated with sources of penetrant at various activities, the solubilities are determined, and a sorption isotherm is generated. Ideal and typical non-ideal sorption isotherms for a non-swelling penetrant in a polymer are shown in Figure 3. A dual sorption model has been utilized extensively to explain the equilibrium sorption data for the type of non-ideal isotherm presented in Figure 3 (4–7). This model postulates that sorption occurs by two

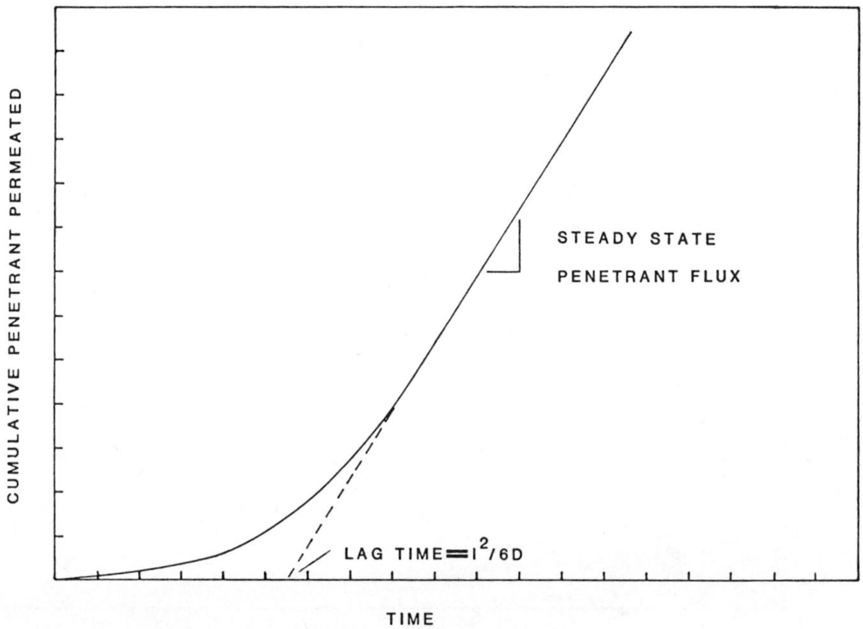

FIGURE 2 Results of a typical membrane permeation experiment.

mechanisms, a simple disolution process involving freely diffusing molecules and an adsorptive process yielding immobilized molecules. Since the immobilized molecules do not contribute to the overall permeation process, this portion of the apparent equilibrium saturation concentration should be factored out of the analysis of the steady state data.

DRUG TRANSPORT IN SKIN

Human skin is a complex structure composed of several layers containing a variety of differentiated cells. The outermost layer of the skin, the stratum corneum, is composed of dead keratinized, partially dessicated epidermal cells. This closely packed cell structure contains about 40% protein, 40% water and 20% lipid. The majority of the resistance to the transport of drug across skin resides in the stratum corneum (2). A diagram showing the structure of human skin is presented as Figure 4.

Scheuplein has shown that the diffusion of drugs across skin is not an active biological process (2). Therefore, the stratum corneum can be

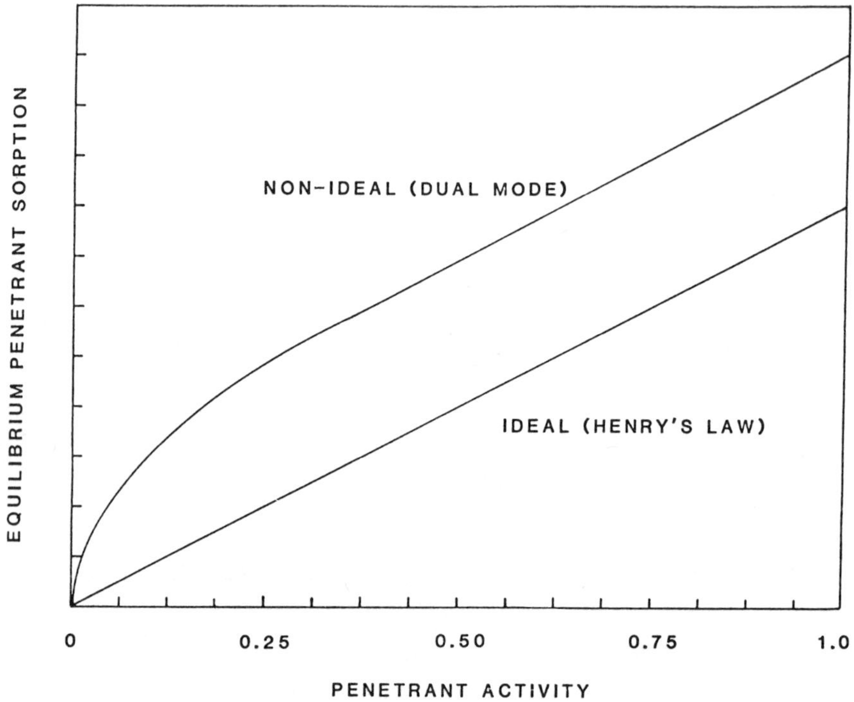

FIGURE 3 Ideal and non-ideal sorption isotherms for non-swelling penetrants.

treated as a polymer membrane in terms of its drug transport characteristics. The permeability of a drug across whole skin or isolated epidermis can be determined *in vitro* in a manner essentially identical to that discussed in the previous section (8). Using a constant unit activity source of drug and an infinitely dilute receptor, the steady state drug flux across the skin can be measured. This is the maximum drug flux at the temperature of interest and is termed the inherent skin drug flux (J_{SKIN}). As with simple membranes, this flux is related to the inherent permeability by the stratum corneum thickness.

$$J_{SKIN} = \frac{\bar{P}}{\ell_{SKIN}} = \frac{DC_S}{\ell_{SKIN}}$$

Both the inherent permeability and thickness of the stratum corneum are known to vary significantly both between individuals and among different body sites for the same individual. These variabilities can contribute to even greater variability in the value of J_{SKIN} (9). The factors

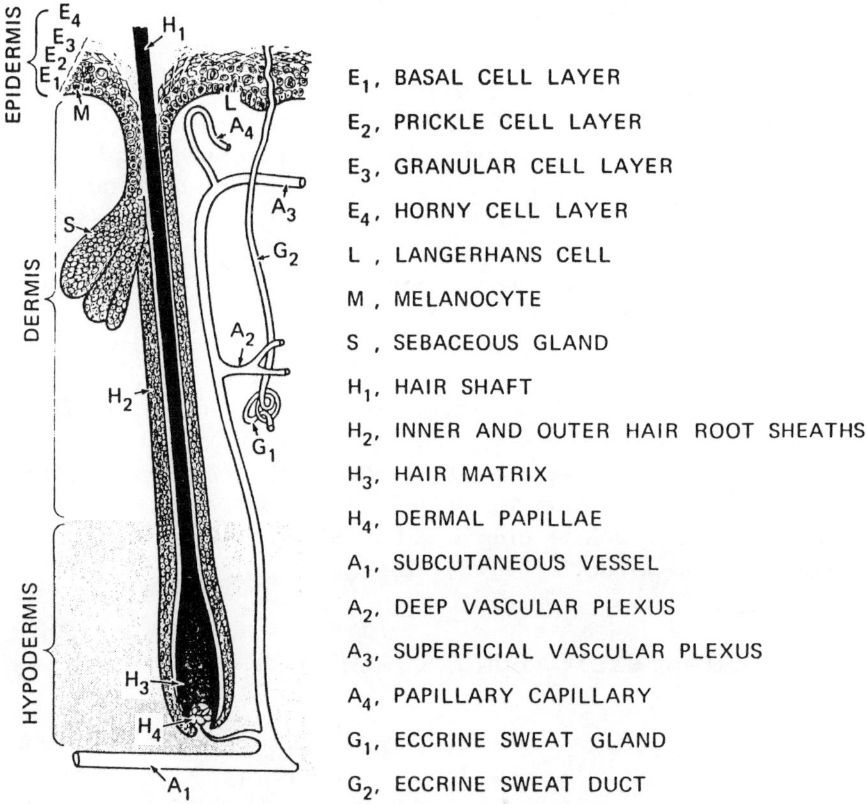

FIGURE 4 Cross-sectional schematic diagram of human skin.

causing these variabilities in skin are not well understood and it is not possible to determine, *a priori*, a value of J_{SKIN} for an individual. Therefore, one of the major goals in the development of a transdermal system is the minimization of the variability in the drug input rate caused by variability in the inherent skin drug flux.

Drug sorption in the stratum corneum is, frequently, non-ideal in nature. In addition to normal dissolution, there are specific binding sites for the drug within the skin. Chandrasekaran, *et al.* have shown that the dual sorption model can be applied successfully to the sorption of drugs in skin (10). Determination of the total drug sorption by the skin and the fraction of bound drug is important in the optimization of a transdermal dosage form.

DRUG TRANSPORT IN SYNTHETIC MEMBRANES

Generally, the rate controlling membrane for a transdermal system is selected to give a steady state drug flux equal to or less than the inherent skin drug flux. The determination of the membrane steady state drug flux is identical to that used for skin. Since the rate control membranes are usually unfilled, rubbery polymers not swollen by the drug, ideal solution behavior is observed. To minimize the mass of drug contained within the membrane, a polymer with a high diffusion coefficient and a low solubility, giving the correct permeability, is chosen.

The drug transport requirements for the contact adhesive layer differ substantially from those of the rate controlling membrane. The steady state drug flux through the adhesive should be sufficiently great relative to that of the membrane so that the adhesive does not increase the resistance to drug release. As with the rate control membrane, a high diffusion coefficient and low drug solubility for the adhesive are preferred. The adhesive may contain drug at saturation or at a loading above the saturation concentration.

TRANSDERMAL SYSTEM FUNCTIONALITY IN VITRO

A typical drug release rate pattern for a rate controlling membrane system is presented in Figure 5. Initially, there is a period of rapid drug release during which drug is desorbed from the contact adhesive and the steady state concentration profile across the membrane and adhesive is established. The size of this initial burst primarily depends on the drug loading of the contact adhesive. After the steady state concentration gradient is established, drug is released from the system at an essentially constant rate for the remainder of the system's intended duration. If the system is evaluated beyond the intended duration, the rate of drug release will begin to fall when the concentration of drug in the reservoir falls below its saturation concentration.

TRANSDERMAL SYSTEM FUNCTIONALITY IN VIVO

At steady state, the net flux of drug from a transdermal delivery system through the skin is controlled by both the skin and the system since the rate control membrane was chosen to have a steady state flux at or below the inherent skin drug flux. The net steady state drug flux can be determined using a series resistance (R) model. Specifically,

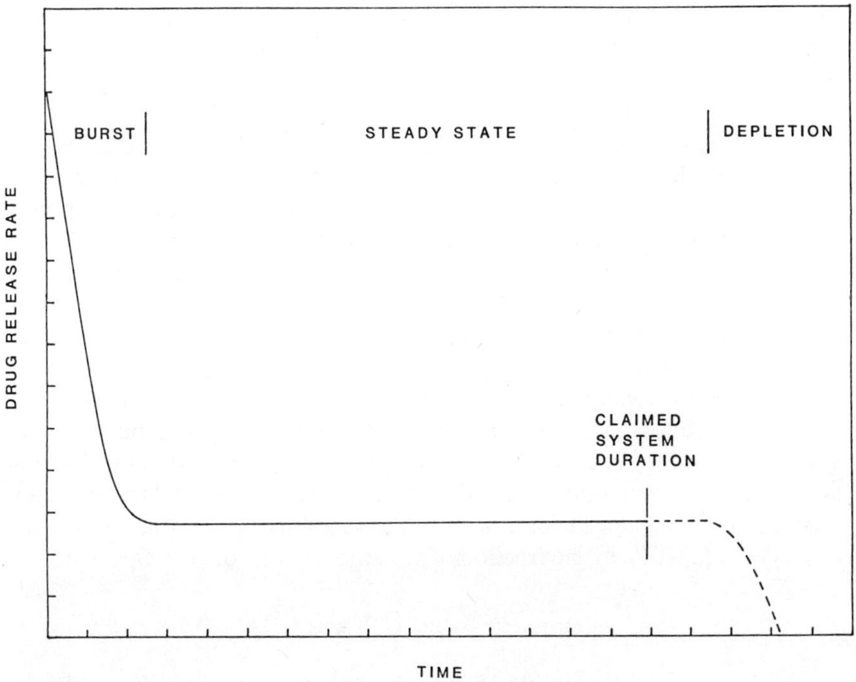

FIGURE 5 Idealized drug release kinetics for a rate-controlling membrane transdermal therapeutic system.

$$R_{NET} = R_{SKIN} + R_{SYSTEM}$$

or, in terms of the steady state drug fluxes,

$$\frac{1}{J_{NET}} = \frac{1}{J_{SKIN}} + \frac{1}{J_{SYSTEM}}$$

where J_{SYSTEM} is the steady state drug release rate for the system, *in vitro*. The fraction of rate control offered by the skin and the system are:

$$f_{SKIN} = \frac{J_{NET}}{J_{SKIN}}$$

and

$$f_{SYSTEM} = \frac{J_{NET}}{J_{SYSTEM}}, \text{ respectively.}$$

When a system initially is applied to the skin there is a transient period during which time the steady state concentration profile within the skin is established. The system must supply sufficient drug to establish this gradient and to saturate the specific binding sites. The functionality of a system providing a high fraction of rate control could be characterized by an extremely long lag time due to the relatively slow rate at which drug is presented to the skin. To overcome this problem, a loading of drug equal to the amount necessary to satisfy the binding sites and establish the gradient is incorporated into the contact adhesive layer. This preload of drug is essentially freely available for absorption by the skin and allows for minimization of the lag time without an adverse effect on the fraction of system rate control.

The rate-controlling membrane system functions in a manner which reduces the effect of variability in the inherent skin drug flux on the variability in the rate of drug input. Assuming the variability in the system drug release rate, at steady state, is small relative to the the skin flux variability, it can be shown that the variability in net drug input rate (Δ_{NET}) is a function of the fraction of system rate control (11). Specifically,

$$\Delta_{NET} = \Delta_{SKIN} (1\text{-}f_{SYSTEM})$$

where Δ_{SKIN} is the variability in the inherent skin drug flux. It is evident from this relationship that the net variability can be reduced to zero by making the system completely rate controlling. However, the increase in system rate control can be accomplished only by reduction of the drug release rate. For a given drug input rate, a decrease in drug release rate necessitates a corresponding increase in system delivery area. The optimization between system size and rate control is central to the development of a transdermal delivery system. This optimization must be performed on a drug by drug basis, taking into account the therapeutic index and dermal toxicity of the drug among other factors.

SUMMARY

A relatively simple approach to the characterization of the transport processes involved in transdermal drug delivery has been presented. The transport of drugs across both synthetic membranes and human skin can be analyzed using well defined models from polymer science. The critical optimization points in the design of a transdermal therapeutic system have been identified and discussed.

REFERENCES

1. S.K. Chandrasekaram and J.E. Shaw, in *Contemporary Topics in Polymer Science*, Pearce and Schaefgen, Eds., 2, 291–308, Plenum, New York (1977).
2. R. Scheuplein and I. Blank, *Physiol. Rev.*, 51, 702 (1971).
3. J. Crank, *The Mathematics of Diffusion*, Oxford University Press, New York (1970).
4. A.R. Assink, *J. Polymer Sci.*, 13, 1665 (1975).
5. A.S. Michaels, W.R. Vieth and J.A. Barrie, *J. Appl. Phys.*, 34, 1 (1963).
6. D.R. Paul, *J. Polymer Sci.*, 7, 1811 (1969).
7. D.R. Paul and D.R. Kemp, *J. Polymer Sci.*, 41, 79 (1973).
8. A.S. Michaels, S.K. Chandrasekaran and J.E. Shaw, *AICHE J.*, 21, 985 (1975).
9. J.E. Shaw and S.K. Chandrasekaran, in *Drug Absorption*, L.F. Prescott and W.S. Nimmo, Eds., 186–193, ADIS Press, New York (1981).
10. S.K. Chandrasekaran, A.S. Michaels, P.S. Campbell, and J.E. Shaw, *AIChE J.*, 22, 828 (1976).
11. J.E. Shaw and F. Theeuwes, Presented at the Second International Conference on Drug Absorption, Edinburgh, Scotland, September 21–23, 1983.

DISCUSSION

ANSWER (Question inaudible) While the system is in use, no. Initially the contact adhesive is loaded at saturation or above, depending on what the design was. Then when you put it on skin, you can say the drug is at saturation. Half of the drug and adhesive sort of falls out and goes outward towards the skin, and you get a steady state concentration break.

QUESTION How much nitroglycerine remains in the reservoir? Does this indicate that only “cheap” drugs could be used in a transdermal application?

ANSWER The Transderm® nitroglycerin system releases about half of its drug content during the 24 hour use period. And yes, what is left behind comes more and more into consideration when you have an expensive drug or if you have a dangerous or abusable drug. You would not want to take an expensive narcotic and deliver five percent and have a patch containing 95% of the drug we sold. Some of the drugs we're developing now are rather expensive, and there is waste in the manufacturing process and there is waste in therapy. But compare it to oral administration where you get 100% of the drug into your mouth, but only get a small fraction from most drugs into the blood stream where it counts. There's an apparent waste there too. It takes a little reeducation of the regulatory authority, as to what bio availability really is. You don't make comparisons based on how much drug comes out of the system and growth.

QUESTION Do you have any idea of the market size of this type of product?

ANSWER I think in the United States, the Nitrosystem®, which accounts for about half of the U.S. market is selling at about 50 million dollars a year. It's big and growing. Alza is getting ready to introduce a system through Boeringer-Ingelheim. It's a 7-day system, for the treatment of hypertension containing chloridine. That will be a small system with 7 days of therapy for an asymptomatic disease, and we hope this will be really big. With hypertension, which probably a lot of people in this room have and half of those aren't aware of it, you don't get any symptoms. When you get symptoms is when you take your medicine; it makes you feel lousy. So compliance is a real problem, because if you feel nothing you're not inspired to take your drug. This will be essentially symptom-less satisfactory therapy for seven days. Hopefully, it can be a big thing. The stratum corneum for every drug we've ever looked at has been rate limiting. We've never seen people talk about the reservoir effect of the drug trapped in the subcutaneous fat layers. The transport across the capillary into the blood stream is not a problem. All those other parts of the skin that I showed you, potentially were rate controlling elements. It turns out that the only one we've ever seen which offers any significant rate control is the stratum corneum itself. We're always careful. We always check that out, and we haven't seen it yet. We've looked at 30 or 40 drugs now.

QUESTION What is the role of sweating and the sweat glands?

ANSWER It turns out that after about 24 hours under occlusion, the sweat glands shut off. So it becomes relatively a non issue. Now to continue, the occlusion of sweat glands makes for potential irritation. You're making a very nice place for bacteria to grow, it's dark and warm, and wet. Having a non-occlusive system would be nice, but it's not an obvious design.

QUESTION In the case of a drug having low skin permeability, would it be possible to have a controlled release of a carrier such as DMSO?

ANSWER Yes, in fact another system will be out on the market soon also developed by Alza that will be the first to incorporate a permeation answer and that will be a system releasing the steroid, estradiol, for peri- and post-menopausal women and that will have ethanol. We control the ethanol, not the drug.

QUESTION You didn't say anything about the membrane in your pack. Are you satisfied with your materials you're using for membranes? Do you need better ones? Or what do you need and what would you like to have?

ANSWER Basically, we're pretty satisfied with the membranes we use. We use the "EVA family". Ethylene, vinylacetate, copolymers. You can vary the acetate content and get a wide range of properties. Polyethylene is a relatively slow membrane and 18/28% vinyl acetate "EVA" is a relatively fast membrane and it varies in a nice monotonic way in between; it makes a nice thing to work with. I'm sure there are many other families of membranes that could be used. Polyethylene ethylacrylate is probably just as good, we've just hit upon one that works. Now we do have specific requirements for other systems and more. I mean this is "x" year old technology, it's not what we're working on today — making systems function like this, we know how to do; beating biological interface problems is the tough thing now. Making a drug that isn't inherently acceptable and adapted to the system is what we're really working on now — slow drugs and irritating drugs.

QUESTION How do you adapt the drug dosage to the patient?

ANSWER Good question! When a doctor initiates transdermal therapy, he has to titrate the patient. Most systems, the nitroglycerin system, and almost every transdermal system from now on whether it be Alza's or anyone else's, are going to come in a range of sizes, so what you do is titrate for effect. You start out with the lowest (smallest) system size in area; that's one of the nice things about transdermal, if you want to double the dose, you double the area. So, you might have three sizes and you start with one, if that's not effective, you try the next size up, and if that's not effective, you try the next size up, and if that's not effective, you go to something else because no drug will treat all patients and no transdermal system will treat all patients.

Membranes in Artificial Organs

M.J. LYSAGHT, D.R. BOGGS AND M.H. TAIMISTO

Materials and Membrane Technology Center, Travenol Laboratories, Inc., Round Lake, IL 60073

The human body employs organic membranes for many of its most important processes; hence, artificial organs, which replace or supplant normal physiologic function, increasingly rely on synthetic membranes to achieve selectivity, permeation, and isolation. By far the largest such application is Hemodialysis which maintains nearly a quarter million patients by thrice weekly treatments in an endeavor that consumes over twenty-million disposable dialysers (Area ~$1M^2$) each year. Next in importance is membrane moderated blood oxygenation during coronary surgery. The technology for both Hemodialysis and Oxygenation encompasses mass transport, materials biocompatibility, and membrane science, and has advanced to a very high plane. Plasmapheresis and Hemofiltration represent recently developed membrane approaches to extracorporeal blood purification. Both treatment modalities are in early stages of clinical application and their technologies are still growing and maturing. Future developments in the field of membranes for artificial organs are likely to include (i) implantable or wearable descendants of present-day intermittent prostheses (ii) hybrid organs in which membranes are combined with cellular clones or immobilized enzymes to augment transport related function with biochemical activity.

INTRODUCTION AND OVERVIEW

The major organs of the human body employ natural membranes for the separation, concentration, and purification of biological fluids. Developers of artificial organs have quite naturally relied upon synthetic membranes for the same purpose. In this review, we will describe the current status and future directions of membrane moderated devices for extracorporeal blood processing.

By far the largest application of membranes in medicine, in fact one of the largest discreet applications of membranes for any purpose, is *hemodialysis* for the treatment of kidney failure. The kidneys are fist-size organs located below the lungs. Each year, through injury or illness, about one person in 10,000 suffers irreversible kidney damage. Until the early 60s, this condition was universally fatal. However, by 1985, the number of anephric patients, i.e., those lacking kidney function, maintained by thrice-weekly four-to-six hour dialysis treatments has reached nearly a quarter million. The membranes that provide this therapy are housed in a hemodialyzer. A hemodialyzer is essentially a disposable 0.6 to $1.5\,M^2$ hollow fiber or stacked plate device, costing \$10–\$20 U.S. dollars that is capable of processing blood continuously at rates of 200–$300\,cm^3/min$. A hemodialyzer performs three functions during each 4–6 hour treatment: it removes about 75 grams of toxins from the blood; it removes 2–8 pounds of water, which the patient has accumulated since his or her last treatment; and it adjusts the plasma concentration of electrolytes and buffers such as sodium or potassium or bicarbonate. Most hemodialysis patients adapt very well physically and psychologically and enjoy lives that are substantially more normal than might at first seem plausible. In the United States alone, the ESRD (End Stage Renal Disease) population is approaching 90,000 patients annually with an average annual federally paid bill of over \$2.8 billion U.S. dollars or \$30,000 per patient a year. Virtually all industrially developed nations in the free world (except Great Britain) provide dialysis therapy to all those requiring it and the worldwide ESRD population continues to grow at a steady pace.

Several alternatives to hemodialysis are available for patients with end stage renal disease. *Transplantation* — the surgical implantation of a donor (or cadaver) organ — is an ideal therapy, especially for young patients, but success is elusive. Despite advances in immunosuppression, the five-year graft survival rate is less than 75%. Moreover, because of the limited number of willing suitable donors, there are simply not enough donor kidneys to go round. In 1985, 7,800 out of 85,000 U.S. patients receive transplants. *CAPD*, continuous ambulatory peritoneal dialysis, involves repeated batch-wise flushings of the peritoneal cavity (belly) with saline solution and removal of uremic toxins by filtration/dialysis from the blood through the body's peritoneal membrane. The process is less costly than hemodialysis and frees the patient from machine dependence and other life-style burdens imposed by the thrice weekly hemodialysis schedule. Its drawbacks include a high infection rate (one infectious episode every 14 months) and the necessity of maintaining a permanent tube-like catheter in the belly. About 16% of non-transplanted ESRD patients in the U.S. elect CAPD over dialysis. This percentage is gradually increasing. Finally, *hemofiltration* is a membrane process that functions by convection rather than diffusion. It is generally regarded as superior to hemodialysis because it offers more complete toxin removal and fewer side effects. However, because hemofiltration is more expensive than hemodialysis, its penetration and acceptance are limited. In Europe, about three percent of all ESRD patients receive hemofiltration rather than CAPD or hemodialysis. In the United States and Japan, the process is used almost exclusively for patients with sudden reversible failure of the kidneys, associated with system trauma, multi-organ failure, surgery, and the like. This form of hemofiltration is termed CAVH (continuous arterio-venous hemofiltration) and is of special technical interest because it requires no external pumps or energy sources and is thus considered a forbearer of wearable or implantible devices.

Blood oxygenators are gas exchange devices that provide the blood with oxygen and remove carbon dioxide. They are used to replace lung function during major coronary surgery when heart and pulmonary circulation are interrutped for one to three hours. Oxygenators exchange about 200 cm^3 (STP)/minute of oxygen and carbon dioxide. They are available in capillary and flat plate format, process from 1.5 to 3 liters of blood per minute, and cost \$300–\$400 U.S. dollars. About 200,000 "open heart" operations are performed annually. Roughly half of these involve membrane oxygenation; the remainder employ bubble oxygenators, which simply sparge oxygen through a blood stream. Use of membrane oxygenation for long term life support in the presence of diseased lungs is the source of ongoing research, but has yet to reach routine clinical practice.

Plasmapheresis is a process for removing plasma proteins. It is well established as a preparative (donor) procedure as a source of such widely utilized biological products as albumin and Factor VIII. Albumin is used to treat traumatic drops in blood pressure; Factor VIII is essential to the treatment of hemophilia. Recently, plasmapheresis has been selectively applied therapeutically to remove proteins from a patient's circulation in auto-immune diseases and other immune globulin derangements. Roughly 12 million preparative (donor) procedures are performed annually. The desired separation is between cellular elements (platelets, white and red blood cells) and the liquid portion of the blood. Centrifugation is the traditional approach, but membranes have been increasingly utilized over the past five years for therapeutic plasmapheresis. Currently, about 100,000 to 200,000 therapeutic plasma exchanges are conducted each year, with membrane representing 10–5% of these treatments. Interestingly, membrane plasmapheresis is much more widely accepted in Japan than in other countries. Despite its relatively modest utilization ($\ll$0.1% of hemodialysis), plasmapheresis is a subject of intense interest and abundant innovation. Some investigators believe it represents the next frontier of artificial organs; others feel that membranes may partially displace centrifugation from the large and lucrative plasma collection market.

These processes are summarized in Table I. Artificial Organ Research represents a hybrid technology and an insular one as well: it tends to have its own journals, societies, meetings, etc. Table II lists information resources for readers interested in pursuing particular therapies in greater detail.

SPECIAL FEATURES OF BIOMEDICAL DEVICES

The requirement for direct contact with circulating human blood places special demands and constraints upon membrane devices used as artificial organs. Devices must be sterile, non-pyrogenic, non-toxic, non-complement activating, and non-hemolytic. These requirements, grouped under the term *biocompatibility* are, in the main, characterized by judgmentalism, empiricism, anectode, and pragmatism. Sterility is the most important compatibility aspect of any blood contacting device.[1] Most membrane devices are sterilized by several hours exposure to moist ethylene oxide vapor. Residuals always represent a concern (and thus an expense) with chemical sterilization; membranes capable of being sterilized by steam autoclaving or irradiation, while desirable, are generally unavailable.

Destruction of living bacteria is assured by sterilization; however, absence of killed bacteria is also required. The receptors of the body's immune system do not in general differentiate dead from live organisms; fever and other host-protection responses are triggered by even nanogram quantities of *pyrogens*, as the fever-provoking constituents of living or dead micro organisms are termed.[2–3] The normal environment is rife with such pyrogens, and extensive precautions are required to produce membrane and devices with levels low enough to be considered safe. Pyrogenic reactions no longer represent a serious risk in medical devices because of accumulated manufacturing experience and because of the development over the past decade of simple, reliable tests to quantify their presence.

Toxicity, which can take many forms remains a concern in devices for medical applications. One of the more puzzling reactions during hemodialysis is the misnomered "first use" or hypersensitivity syndrome, a serious allergic-type reaction that occurs in about 3–5 of every 100,000 treatments. (4) Its origins are multi-factorial and, at present, etiology methods for prevention are not known. Other sources of toxicity are better understood. Most solvents, pore formers, etc., used in fabricating membranes (glycerin excepted) are toxic to some degree when introduced directly into blood. (1) So a primary requirement of membranes for blood contact is absence of such leachables or extractables. Antioxidants, surfactants (for membranes made from hydrophobic polymers) and the like can be particularly difficult to remove and in fact are sometimes

TABLE I

Overview of membrane moderated artificial organs

	Scope				Device				
	Application	Function	Patients per year	Treatments per year	Membranes	Area M^2	Blood flow ml/min	Filtrate or Dialysate ml/min	Device cost $/US
Hemodialysis	Chronic Renal Failure	4–6 hrs. 3 time/week — Removes 25 grams toxin — Removes 2–8 lbs. fluid — Adjusts plasma Electrolytes	225,000	35,000,000	Predominantly homogeneous cellulosic capillaries or sheets	0.7–1.4	200–300	500	$10–20
Intermittant Hemofiltration			3,000	500,000	Protein retentive anisotropic ultrafiltration membranes; Capillary or sheet.	0.5–1.5	300–400	110–150	approx. $50
Continuous Hemofiltration (CAVH)	Acute Renal Failure and Oedema	Removes 600 cc fluid plus 1 gram toxins per hour. Process is continuous 24 hours/ day. Usually employs no blood pump.	8,000	40,000		0.2–0.5	50–100	10	approx. $70

TABLE I (Cont'd)

Overview of membrane moderated artificial organs

	Scope				Device				
	Application	Function	Patients per year	Treatments per year	Membranes	Area M^2	Blood flow ml/min	Filtrate or Dialysate ml/min	Device cost $/US
Oxygenation	Replaces cardio pulmonary function during heart surgery	Delivers 0.2 L of O_2 minute; removes equivalent CO_2 Lasts 1–4 hours.	160,000	16,000	Hydrophobic fluorocarbon or olefinic microporous capillaries or sheets.	1.0–3.0	2,000–40,000	DNA	$200–400
Plasma Filtration	Primarily for for autoimmune disease	Typically applied daily during fulminant or crisis stage of diseae Removes 100–150 gram of protein	<3,000	<25,000	Olefinic microporous hollow fibers	0.1–0.3	70–100	30	approx. $75

Note: Cost, populations are approximate and refer to 1985.

TABLE II

Artificial Organ Literature

Hemodialysis:
Colton, CK and Lowrie, EG. Hemodialysis: Physical Principles and Technical Considerations. In: The Kidney (Brenner, BM and Rector, FC, eds), Saunders, Philadelphia, 1981, pp. 2425–2487.

Drukker, W, Parsons, FM, and Maher, JF. Replacement of Renal Function by Dialysis. Martinus Nijoff, The Hague, 1979.

Frost, JH. Technical Aspects of Renal Dialysis, Pitman Medical, Kent, (Eng.) 1976.

Hemofiltration
Henderson, LW, Quellhorst, EW, Baldamus, CA, Lysaght, MJ. Hemofiltration, Springer-Verlag, Heidelberg, 1985.

Schaeffer, K, Koch, KM, Quellhorst EW, Von Heurath, D. Hemofiltration as Vol. 32 Contr. Nephrol. Kauger Basel, 1982.

CAVH
Kramer, P. Arterio-venous Hemofiltration. Springer-Verlag Heidelberg, 1985.

Plasmapheresis
Lysaght, MJ and Gurland, HJ. Plasma Separation and Plasma Fractionation: Current Status and Future Directions. Karger Bagle, 1983.

MacPherson JL and Kasprisin, DO. Therapeutic Hemapheresis (Vols. I and II) CRS Press, Boca Raton, Florida, 1985.

Nose Y, Malchesky, P, Smith JW, Krakaver, RS. Plasmapheresis: Therapeutic Applications and New Techniques. Raven Press, New York, 1983.

Oxygenation:
Ionescu, MJ. Techniques in Extracorporeal Circulation. Butterworth, London, 1981.

Kenedi RM, Artificial Organs University Park Press, Strathelyde, 1977.

Journals: Blood Purification (Karger)
Artificial Organs (Raven Press)
Dialysis and Transplantation (Creative Age)
Life Support Systems (Saunders)
Plasma Therapy (Laux)

Annual Transactions: Transactions of the American Society of Artificial Internal Organs, ASAIO Press

Proceedings of the European Dialysis Transplant Association — Pitman

required for proper membrane functioning. The intrinsic material of membrane can also be toxic. Hydrophobic polymers tend to absorb and denature several layers of protein, altering the polymer's transport property and requiring more aggressive anticoagulation measures. (5) Most hydrophilic materials, notably those that are biologically derived, contain –OH, –COOH, C=O surface groups in quantities capable of being recognized by the body's seveal immune systems. A dramatic example of this is cellulosic dialysis membrane, which activates the body's "complement system" (6, 7). The complement system is a quasi-immune regulator; the biological sequelae of activation are formidable, but the only frank clinical consequence is a transitory (20–30 minute) sequestration of the white blood cells in the lungs. Nevertheless, any non-physiologic response is worrisome, and increasing effort is being devoted to modifying blood contacting membrane to eliminate this reaction. Materials biocompatibility then may be viewed as a negative attribute; a material is biocompatible to the extent that it contains no infectious or pyrogenic materials and no untoward leachables or extractables, and to the extent that its intrinsic structure is not recognized as foreign by the body's many defense mechanisms.

Fluid mechanic constraints also attend the design of biomedical devices. Flow is almost always laminar and in fact very low Reynolds numbers ("creeping flow") are common. Wall shear rates must be above approximately $100\,sec^{-1}$ to maintain the Newtonian character of blood and below about $1500–3000\,sec^{-1}$ to avoid red cell damage. (8) In devices employing microporous membranes, allowable shear rates are further constrained by fear of cell lysis (rupture) by the similarly sized pores. (9) In general, abrupt expansions or contractions, sharp ridges, separated flow, and local stagnation are to be avoided because of their tendency to promote hemolysis or clotting.

In some ways, the design of artificial organs is actually easier than that of general membranes. Extracorporeal circuits generally operate at less than 500 mmHg (10 PSI) transmembrane pressure and almost never at temperatures above 37°C. They are rarely used for more than 24 hours, usually less. They are not exposed to acids, bases, or corrosive materials. They are utilized in a narrow and rigorously defined protocol. Since the devices are disposable, very large quantities are produced (breakeven points for new designs are in the hundreds of thousands of units) and the manufacturing learning curve may proceed to very advanced levels.

TRANSPORT IN ARTIFICIAL ORGANS

Except for the special considerations discussed above, artificial organs resemble conventional mass exchange devices in the structure and properties of their membranes and in the principles that govern their use.

Dialysis — A typical circuit for hemodialysis is shown in Figure 1a. Blood flow is 200 ml/min; dialysate, which is counter current, is 500 ml/

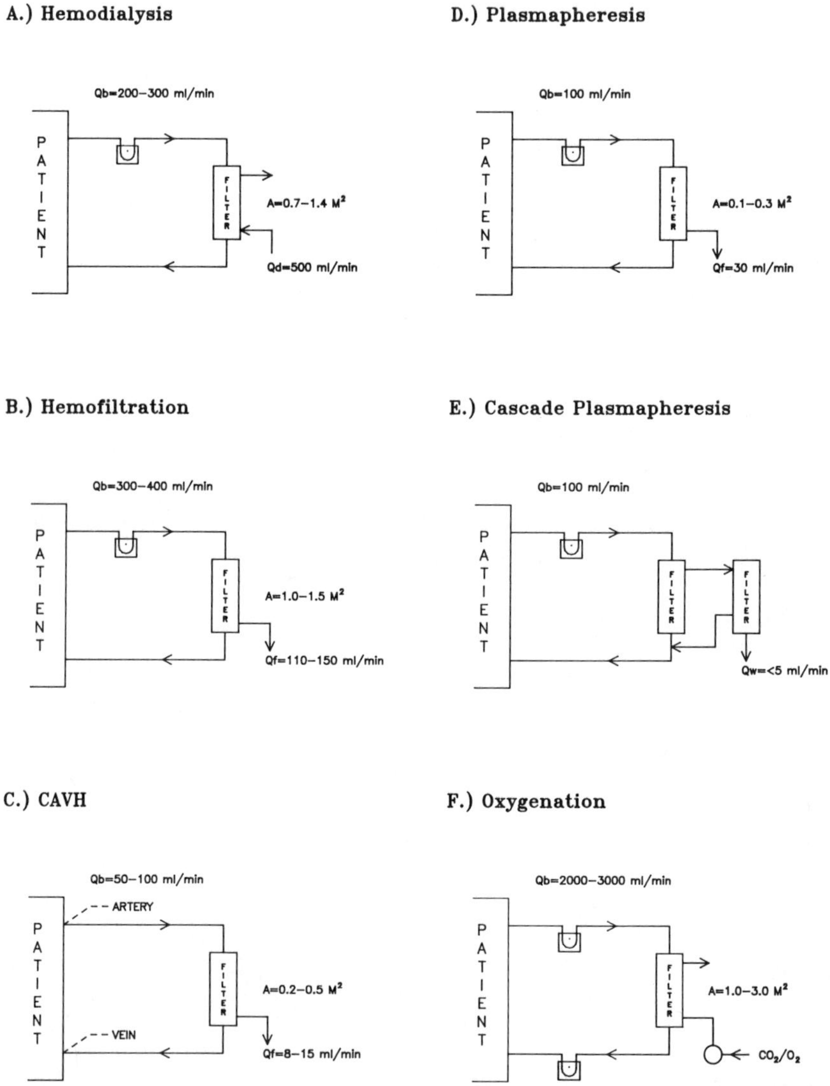

FIGURE 1

min. Devices may employ hollow fiber or flat sheet membrane; area varies from 0.6 to approximately 1.5 M^2. Membranes are most often Cuprophan, a thin (circa 10 micron) cellulosic manufactured by the cuprammonium process by ENKA, Wuppertal, Germany. (10) Typical performance is summarized in Figure 2. Solute removal is universally characterized by clearance, the virtual volume of a blood cleared of a given solute per unit time.

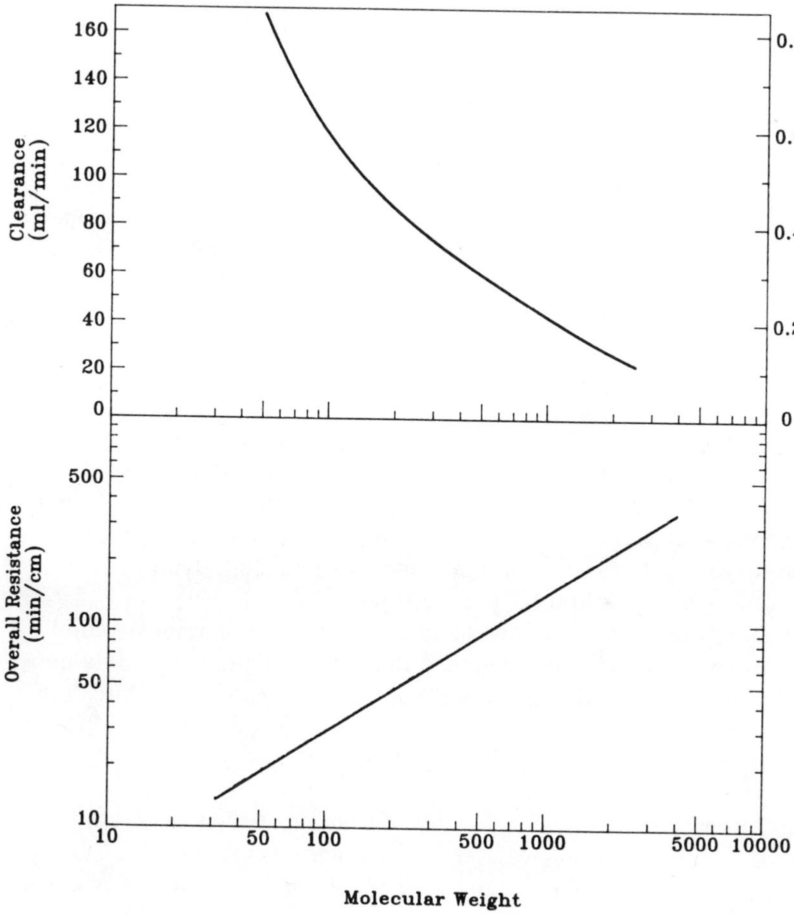

FIGURE 2

$$Cl = \frac{Q_B(C_i - C_o)}{C_i} \tag{1}$$

Where Cl = Clearance in ml/min; Q_B = inlet blood flow in ml/min, c = concentration, and subscripts i, o refer to inlet and outlet.

Clearance is plotted as the left y-axis on Figure 2, the right axis shows the extraction ratio or the clearance divided by inlet blood flow rate. Approximately 80% of incoming urea is "cleared" from the blood; both clearance and extraction coefficient decrease with increasing molecular weight (larger molecules diffuse more slowly than smaller molecules) until about 5000 Daltons where fractional removal falls to between 5 and

10%. Of course, solute transport can be described by more conventional engineering expressions, i.e.:

$$J = K_O A C \tag{2}$$

where J = solute mass flow rate in g/min; K_O is a mass overall transfer coefficient in cm/min, A = device area in cm^3, and C is the logmean concentration difference in g/cm^3 for the solute being transported. K_O is conveniently divided into its blood (K_B) membrane, (K_M) and dialysate (K_D) components which are reciprocally additive:

$$1/K_O = 1/K_B + 1/K_M + 1/K_D \tag{3}$$

The parameters are plotted (as reciprocals) in the bottom panel of Figure 2. Rather sophisticated theory based upon the classic Graetz solution is available for predicting mass transport from diffusivity, geometry, and flow conditions (11, 12). In practice, simple empirical rules are more widely followed. Design principles and underlying theory are nicely reviewed by Colton (13) and Sargent (14).

In addition to diffusive toxin clearance, hemodialysis removes fluid by convection at a rate that is rarely less than 10 cc/minute and rarely more than 20 cc/minute. Fluid removal follows the simplest form of Darcy's law

$$J_F = KA\,TMP \tag{4}$$

J_F = filtration rate in ml/min, K = ultrafiltration coefficient in ml/(min-M^2-mmHg); A = area in M^2, TMP = average membrane pressure in mmHg. The transmembrane pressure is usually established via the dialysate circuit and is often called negative pressure. TMP ranges between 100 and 500 mmHg. In practice, its selection, which is guided by the rating of the dialyzer, controls fluid loss. Hence, reproducibility of ultrafiltration coefficient is an important membrane/device characteristic. Filtration influences toxin removal. A rather satisfactory theory for this coupled transport exists (15), but because its magnitude is small, this effect is usually conveniently ignored. Approximately 85% of the world's dialysis membrane is Cuprophan brand regenerated cellulose. Cuprophan is supplied films or fibers containing from under 10% glycerine; it expands approximately two fold upon wetting. The resistance of cuprophane membranes ($1/K_M$) has decreased constantly over the past 15 years; for example, urea resistance has fallen from over 20 min/cm in the early 70's to under 10 min/m today. Cost is approximately \$.50 to \$1.50/M^2 for flat sheet and \$2.50 to \$4.00/M^2 for hollow fibers. The large reliance upon

Cuprophan, despite an abundance of potential competitors, derives from its reliability, its constant improvement in performance, and its high mechanical strength. The remaining 15% of dialyzers are made from a wide variety of cellulose non-cellulosic membranes. A complete listing of available dialyzers and membrane has been completed by Sigdell. (16)

Hemofiltration — Hemofiltration, an alternative process to dialysis, is more often used in acute (17) rather than chronic (18) renal failure. The acute circuit — notable for absence of a blood pump and reliance upon existing arterio-venous pressure differentials — is shown in Figure 1c; a typical chronic circuit is shown in Figure 1b. In both formats, clearance is approximately numerically equal to ultrafiltration rate; only the chronic circuit normally requires replacement of lost volume with parenteral fluid. In the acute circuit, filtration rate per unit area is between 5 and 15 cm^3/minute and is usually too low to be in the plateau region of concentration polarization. Filtration results from both physiologic pressure and the siphon effect of the collection reservoir; in combination these rarely exceed 100 mmHg. Osmotic pressure plays an important and somewhat limiting role in controlling filtration. (19, 20) Blood flow is a dependent variable and varies with AV pressure difference and device geometry.

In chronic hemofiltration, clearance values equivalent to those of dialysis are desired. This requires a blood flow rate of 300–400 ml/minute; all operation is in the maximum (asymptotic) range of ultrafiltration rate. Filtration rate is governed by concentration polarization of plasma proteins and is influenced by conditions of geometry and flow:

$$\frac{Q}{A} = K\left(\frac{\gamma W}{L}\right)^{0.5} \ln\left(\frac{95}{C/W}\right) \tag{5}$$

where Q = filtrate flow rate in ml/min, A = area, K = 2.28 for blood with an inlet hematocrit of 20%; γW = wall shear rate, sec-1, L = length in cm; C/W = inlet or average bulk protein concentration in g/100 cm^3.

The empirical and theoretical bases of the equation is described by Colton (21), Ofsthun (22) and Lysaght (23). Hemofiltration membranes were first introduced by Amicon as copolymers of PAN and PVC. Interestingly, their development by Ford and Paulson represented the first achievement of the anisotropic phase-inversion membrane in hollow fiber form (24). The most widely used membranes today are polysulfone (Amicon) polyamide (Ganbro), PAN (ASAHI Hospal), and cellulose triacetate (Sarturiuos). The microstructure and transport characteristics membranes differ little from their counterparts in industrial and laboratory ultrafiltration.

Oxygenation — Oxygenation is essentially a gas-liquid exchange process that takes place across or through a barrier membrane (25, 26). Oxygen is transferred into the blood; CO_2 is removed from it. The advantage of membrane devices, in comparison with direct-contact gas liquid exchangers (bubble oxygenators), is reduced trauma to the blood; this is regarded as especially important during long (>1–2 hours) periods of bypass circulation. Devices vary in area from 1 to 5 M^2 and transfer up to 200 cm^3/min of gas to/from blood. Devices originally employed silicone or silicone polycarbonate membranes; these have more recently been supplemented with hydrophobic (teflon or polyolefin) microporous sheets or fibers. Such membranes are extremely permeable to gaseous species and essentially all the resistance to transport occurs in the blood film adjacent to the membrane surface. Being more soluble, CO_2 is transported through this liquid phase more readily than oxygen and, in practice, some CO_2 is added to the sweep stream to avoid excess CO_2 removal. Although several ingenious designs have been proposed that incorporate secondary flow to minimize blood side resistance, existing filters all employ conventional flat sheet or hollow fiber geometries. The rate of gas transport increases with blood flow rate, with hematocrit, and with the difference in oxygen activity between the liquid and gas phase. The latter depends upon the degree of saturation of the incoming blood. The classical (27) transport equation is:

$$V_{O_2} = Q_B\, 4 \times 10^{-5} H(S_i - S_o) + g_{O_2}(P_o - P_i)$$

where V_{O_2} = oxygen transfer rate in ml (STP)/min, Q_B is inlet blood flow rate in ml/min, S_i and S_o are the fractional blood saturations of the device inlet and outlet, P_o and P_i are the inlet and outlet partial oxygen pressures, mmHg and g_{O_2} is the solubility coefficient of oxygen in blood = .67H + .947 (1-H).

Device performance is usually described by its *rated* blood flow in ml/min, an arbitrary but helpful parameter. Rated blood flow is the highest inlet blood flow rate that yields a 95% outlet saturation with blood under standard operating and inlet conditions (hemoglobin, 15 g/100 ml blood; P_{O_2}, 40 mmHg, inlet saturation, 65%: O_2 content; 15 ml/100 ml). At lower inlet flow rates, outlet saturation will be higher (often 100%); at higher inlet flow rates, outlet saturation decreases. Most membrane oxygenators have rated flows of 2–3 liters/minute; bubble oxygenators have higher ratings.

Plasmapheresis — Plasmapheresis shares many broad similarities with hemofiltration. The former yields plasma as filtrate; the latter yields protein-free electrolyte. Other major differences are generally a matter of scale. Refer to Figure 1e. Blood and filtrate flow rates are 100 and 30 cm^3/min respectively. Filtrate is replaced with colloid solutions (usually 5% albumin) and the tolerance of the patient to this fluid — rather than technical factors — establishes the upper limit on filtration rate. Contemporary membrane separators have a sieving coefficient of 1.0 for all plasma proteins (including the large lipoproteins), and thus the filtrate is equal in composition to controls obtained by centrifugation. (28) Most plasma exchange sessions last 1–2 hours and exchange a total volume of 2–3 liters.

Transport during filtration is governed by a concentration boundary layer of red cells, and possibly platelets, rejected by the membrane. (9, 29, 30) This layer is dynamic and its resistance (or ability to limit flow) changes in response to shear rate, pressure, and red cell concentration. Filtration rate is given by the following semi-empirical expression, which should be compared to equation 5.

$$\frac{Q_F}{A} = 0.07\frac{(a^4)^{1/3}}{L}\ln\frac{(.95)}{H_i} \tag{7}$$

where a = red cell diameter (aproximately 4 microns; L = length cm; H = bulk hematocrit, W, Q_F, and A are as defined previously.

The rupture of red cells, with subsequent release of their contents, was more often a problem associated with the development of plasmapheresis than with other processes. Such hemolysis is now known (9, 29) to occur a above certain critical transmembrane pressure, which itself varies in quasi-empirical fashion with the product of pore radius and wall shear rate.

The exponent describing the dependence filtration rate upon shear is closer to unity in plasmafiltration than in hemofiltration (compare equations 4 and 6). One rather curious consequence is that filtration rate stays approximately constant as the number of fibers (or plates) in a filter are decreased, as the loss in area is balanced by a directly proportional increase in shear rate. Progress in design of membrane plasmafilters has been very marked. Existing hollow fiber devices have only 20–40% of the surface area of earlier prototype and although they employ very similar membranes, are nonetheless able to offer equivalent performance (30). Moreover, two versions incorporate very special design features. The COBE system (31) comprises a parallel channel (stacked plate) configuration whose gap-thickness is adjusted — via a feedback loop and hydraulic clamp — in response to sensed conditions of pressure and flow. As a consequence, this system achieves higher filtration fractions than fixed geometry alternatives. Hemascience (32) has introduced a design incorporating rotating concentric cylinder geometry. The membrane forms the inner (rotating) cylinder; the outer cylinder is stationary. Red cell transport from the membrane is very efficient, probably due to Taylor vortices. The cell achieves filtration rates with 50 cm^2 of membrane equivalent to those of conventional design that employ several thousand cm^2 of membrane surface area. It remains to seen, of course, whether the increased cost implicit in these complicated designs can be justified by their higher efficiency.

Membranes for plasmapheresis are fairly conventional microporous structures with 0.1 to 0.5 micron pore radii. Flat sheet devices incorporate Nucleopore track etch as well as films produced by phase inversion and thermal precipitation. The most widely employed membrane is presently Plasmaphan from Enka, which is prepared in hollow fiber format from polyolefin by thermal precipitation.

Finally, several attempts have been made to utilize membranes to further fractionate the plasma filtrate into an albumin-rich portion for return to the patient and albumin-poor globulin-rich aliquot which is discarded. This approach avoids the cost and other disadvantages of donor-source albumin replacement fluid. The filtration is sometimes accompanied by chilling ("cryofiltration"). Both autologous protocols are used clinically in special cases; neither is yet widely employed (33).

FUTURE TRENDS

Hemodialysis is in its latter growth stages and future developments are likely to be more evolutionary than radical. Because of the widespread use of the therapy, cost considerations will continue to dominate. Devices will therefore become smaller and more efficient. Treatment time is likely to be reduced from the current level of three to four hours to two hours or less. Low cost cellulosic membranes whose structure has been modified to avoid complement activation will be available and will be widely used. Despite this technical progress, the general utilization of hemodialysis will decline partially in response to societal disinclination to provide resources for this purpose and partially because of increased reliance upon transplantation and CAPD.

In contrast, the technology for hemofiltration, plasma filters, and oxygenators is likely to undergo major expansion and evolution over the next decade. Our understanding of the processes, particularly the relevant transport theory, should expand considerably. The detailed mechanisms governing transport should supplant the current black box approach to concentration polarization modeling, and exact numerical treatments will replace the use of device-averaged coefficients. Transport theory based upon more explicit understanding and more precise mathematical descriptions will not only allow the development of more efficient devices at lower costs, but will also encourage exploration of new device formats, particularly those in which the shear forces necessary to minimize polarization are obtained not just from conditions of geometry and flow, but from secondary flow and from the rotation of membranes itself. Thus, over the next decade, a variety of low area (50 to 500 cm^2) devices will become available for these convective processes and will replace the larger areas (5,000 to 10,000 cm^2) that are standard today. It is also likely that the minimization of boundary layer effects implicit in these designs will allow more precise fractionation of albumin from the higher molecular weight globulins, thus rendering feasible cascade filtration and allied processes.

Further progress in the development of "hybrid artificial organs" is also expected (34). Hybrid organs utilize membranes to separate allograft cells from the donor's body. They are currently being investigated for the purpose of creating an artificial pancreas. Cells grow in the interstitial region between the fibers; blood flows though the fiber lumen. Nutrients and oxygen diffuse from the blood to the cells along with glucose; metabolic waste products and insulin are released by the cells and diffuse back into the body. Importantly, the membranes prevent the passage of higher molecular weight immuno active compounds (IgG, IgM) and thus the cells are not "rejected" by the host. Such hybrid organs were originally proposed in 1976 (35), but their successful deployment is extremely complex and their progress is in the early research stage. When fully developed, this approach represents a viable avenue just not to the pancreas but also to the development of prostheses for the liver and various endocrine organs, and a replacement for the synthetic function of the human kidney.

Conventional artificial organs will play an important and expanding role in critical care. Over the next decade, major progress is expected in the application of high-tech membrane devices to support extremely premature neonates and extremely ill, acute patients. In many ways, such temporary, extremely interventionist applications seem more suited to extracorporeal artificial organs than to long-term maintenance applications. For long term use, the goal will shift to the development of implantable or wearable devices connected into the patient's normal circulation and capable of performing on a continuous routine basis the function of the organ they replace.

REFERENCES

1. Klein E. Villaroel D. Evaluation of Hemodialyzers and Dialysis Membranes. HEW Publication #NIH 77-1294 (Stock No. 017-045-00079), 1977, p. 111–129
2. Dinarello CA. Pathogenesis of Fever during Hemodialysis. Contr Nephrol 36: 6-50 (1983)
3. Dinarello CA. The Biology of Interleukin-1 and its Relevance to Hemodialysis. Blood Purification 1: 231–241 (1983)
4. ASAIO Journal 8(3):268–395 (1984)
5. Cooper SL. Peppas NA. Biomaterials: Interfacial Phenomenona and applications. Advances in Chemistry: 199. American Chemical Soceity, Washington, DC. 1982
6. Craddock PR. Complement and Granulocyte Activation and Deactivation during Hemodialysis. In: Plasma Separation and Plasma Fractionation. Eds: Lysaght MJ, Gurland HJ. Karger, Basle, 1983. p. 14–24
7. Chenoweth DE, Cheung AK, Henderson LW. Anaphalotoxin Formation during Hemodialysis: Effects of Different Membranes. Kidney Int 24: 764–769 1983)
8. Lightfoot EN. Transport Phenomena and Living Systems. Wiley Interscience, New York, 1974
9. Zydney AL, Colton CK. Continuous Flow Membrane Plasmapheresis: Theoretical Models for Flux and Hemolysis Predictions. Trans Am Soc Artif Intern Organs 28:408–412 (1982)
10. Baurmeister UR, Pelger M. Tretzel J. Henne W. Development and Properties of Cuprophan Membranes. ENKA AG (privately printed), Suppertal Germany, 1985
11. Davis HR, Parkinson GV. Mass Transfer from Small Capillaries with Wall Resistance in the Laminar Flow Regime. Appl Sci Res 22:20 (1970)
12. Wolf L. Zaltsman S. Optimum Geometry for Artificial Kidney Dialyzers Chem Eng Prog Symp Series 64(84):104–111 (1968)
13. Colton CK, Lowrie EG. Hemodialysis: Physical Principles and Technical Considerations. In: The Kidney (2nd Ed). eds: Brenner BM, Rector FC Saunders, Philadelphia, 1981. p. 2425–2489
14. Sargent JA, Gotch FA. Principles and Biophysics of Dialysis. In: Replacement of Renal Function by Dialysis, eds: Drucker W, Parsons FM. Maher JF. Martinus Nijhoff, The Hague, 1979, p. 38–68
15. Schmidt M. Hemodiafiltration. In: Hemofiltration eds Henderson LW Quellhorst EA, Baldamus CA, Lysaght MJ. Springer Verlag, Heidelberg IN Press (1986)
16. Sigdell JE. Comparison of Hollow Fiber Dialyzers. Artif Organs 5(4): 402–409 (1981); 6(1):77–78 (1982); 6(4):460–462 (1982); 7(3):374–375 (1983); 8(2):233–236 (1984)
17. Kramer P. Continuous Arterio-Venous Hemofiltration. Springer Verlag Heidelberg, 1986
18. Henderson LW, Quellhorst EA, Baldamus CA, Lysaght MJ. Hemofiltration. Springer Verlag, Heidelberg, 1986 (In Press)
19. Lauer A, Saccaggi A, Ronco C, Belledonne M. Glabman S. Bosch J Continuous Arteriovenous Hemofiltration in the Critically Ill Patient. Annals Internal Medicine 99: 455–469 (1983)
20. Lysaght MJ, Schmidt B, Gurland HJ. Filtration Rates and Pressure Driving Force in AV Filtration. Blood Purification 1(3):178–183 (1983)
21. Colton CK, Henderson LW, Ford CA, Lysaght MJ. Kinetics of Hemodiafiltration I. In Vitro Transport Characteristics of a Hollow-fiber Blood Ultrafilter. J Lab Clin Med 85(3):355–371 (1975)
22. Ofsthun NJ, Colton CK, Lysaght MJ. Determinants of Fluid and Solute Removal Rates During Hemofiltration. In: Hemofiltration eds Henderson LW, Quellhorst EA, Baldamus CA, Lysaght MJ. Springer Verlag, Heidelberg. In Press (1986)
23. Lysaght MJ, Ford CA, Colton CK, Stone RA, Henderson LW. Mass Transfer in Clinical Blood Ultrafiltration Devices. In: Technical Aspects of Renal Dialysis ed: Frost TH. Pitman Medical, Kent England, 1978. pp. 81–95

24. Henderson LW. The Beginnings of Hemofiltration. Contr Nephrol 32:1–19 (1982)
25. Ionescu MI. Techniques in Extracorporeal Circulation. 2nd ed Butterworths, London, 1981
26. Kenedi RM. Artificial Organs (Proceedings Strathclyde Bioengineering Seminar) University Park Press, Strathclyde, 1977. Chptrs 1–8.
27. Galletti PM, Richardson PD, Snider MT, Friedman LI. A Standardized Method for Defining the Overall Gas Transfer Performance of Artificial Lungs. Trans Amer Soc Artif Int Organs 18:359–368 (1972)
28. Gurland HJ, Lysaght MJ, Samtleben W. Schmidt B. Comparative Evaluation of Filters Used in Membrane Plasmapheresis. Nephron 36:173–182 (1984)
29. Zydney AL. Cross Flow Membrane Plasmapheresis: An Analysis of Flux and Hemolysis. Doctoral Thesis, MIT, Cambridge MA. (February 1985)
30. Lysaght MJ, Schmidt B, Samtleben W, Gurland HJ. Transport Considerations in Flat Sheet Microporous Membrane Plasmapheresis Plasma Ther Transfus Technol 4:373–385 (1983)
31. Buffaloe GW. Erickson RR, Dau PC. Evaluation of a Parallel Plate Membrane System. J Clin Apheresis 1:86–94 (1983)
32. Fischel H. European Patent Application 83307577.3
33. Lysaght MJ, Samtelben W, Schmidt B, Gurland HJ. Closed Loop Plasmapheresis. In Therapeutic Hemapheresis eds: MacPherson JL, Kasprisin DO. CRC Press, Boca Raton, 1985
34. Galetti PM, Hori M, Sharp DW, Stanley C. Hybrid Artificial Organs (Panel Conference). Trans Am Soc Artif Intern Organs 28:639–646 (1982)
35. Chick WL, PERNA JJ, Lauris V, Low D, Galetti PM, Panol G, Like AA, Whittemore AD, Colton CK, Lysaght MJ. Artificial Pancreas Using Living Beta Cells: Effects on Glucose Homeostasis in Diabetic Rats. Science 197:780–781 (1977)

DISCUSSION

QUESTION What is the cost differential between hemofiltration and dialysis and what can be done to lower it?

ANSWER The major part of the costs for each is the cost of the fluids, and this will favor hemofiltration in time. The cost is about $100 per treatment, and the barrier that's going to have to be overcome will be a self-generation of fluid using membrane-type processes to bring the costs down. That has been a theory proved, and I think that we are going to see this happen, but ever so slowly over the next four or five years.

Functionality, Performance and Applications of Ion-Exchange Membranes

J. J. LEDDY

Chlor-Alkali Technology Center, The Dow Chemical Co.

The exponential growth in development and proven applications of ion-exchange membranes over the last thirty years is classic. Indeed, the availability of perfluorinated membranes, coupled with dimensionally stable electrocatalytic anodes and cathodes, is revolutionizing a major world industry devoted to the production of chlorine and caustic. In that area, the progress in the last ten years has been nothing short of phenomenal. The cell design options created by the availability of high performance membranes have lowered the electrical energy consumption per unit of cell product to very attractive levels. Increased current densities at these high performance levels also make the capital investment more attractive. As well, the purity and high concentration of membrane caustic cell effluent lend high incentive. Spurred initially by environmental concerns surrounding mercury cell technology and edicts by the government of Japan to replace that technology, the ion exchange membrane cell has now surpassed those initiatives and has begun to impact diaphragm cell technology. But for the high cost of capital and an overbuilt industry the commitment to convert to this technology would be greater than it now is, which is 2.7 million metric tons of chlorine and caustic per year, or about 3% of total world capacity. This is not insignificant in a $12 billion/year global industry.

Other applications of various ion-exchange membranes are in Donnan dialysis (metal ion recovery from aqueous industrial wastes), electrolytic regeneration of chromium-bearing solutions (acid etching and stripping operations), fuel cells, HCl electrolysis, water electrolysis, production of potassium stannate, electro-hydrodimerization (adiponitrile), and electrodialysis (especially univalent-selective membranes used in desalination). Battery separator applications are covered in a separate paper.

INTRODUCTION

The first practical synthetic ion exchange membranes permselective, respectively, to anions and cations, were developed in the late 1940's (1,2). The practicality of such membranes hinges as much on their physical strength, chemical stability in a variety of media, and low electrical resistance as on their permselectivity. Such membranes were produced both heterogeneously by dispersion of ion exchange resin in a plastic matrix, and homogeneously by condensation reactions or chemical treatment of graft copolymers (3, 4, 5, 6). Compared to the present day membranes these early membranes were relatively fragile, with poor resistance to attack by oxidizing media. Meares reviews these trends in membrane technology in a recent monograph edited by Flett (7).

Much of the early work was directed toward development of membranes for applications in electrodialysis (8), especially the desalination of brackish water and the production of table salt (9). The early interest in these applications by the Japanese in view of their lack of geologic salt strata, the paucity of fresh water and plenty of sea water inspired them to early successes in the field. In the electrodialysis of sea water the formation of chlorine and caustic at the anode and cathode, respectively, of the dialysis stack is a nuisance. Early practitioners (10, 11, 12) capitalized on the notion of using the membranes in electrolytic cells dedicated to the production of chlorine and caustic. The Japanese were soon to be inspired further by the insistence of their government that their islands be rid in time, of their heavy commitment to mercury cells. Enter a nation of goal-oriented people to such a cause and an upward thrust begins on the exponential curve of technological developments in ion exchange membranes toward implementation in the huge chlor-alkali market (13, 14). That thrust would have lost momentum but for the basic work being done in the laboratories of the duPont Company (15, 16) toward the production of perfluorinated cation exchange membranes.

The chemical nature of the polymers and pendant ion exchange groups in a membrane determines its serviceability, which is also impacted by the degree of homogeneity. A heterogeneous membrane requires a high loading of ion exchange resin particles within the binder phase in order to ensure uninterrupted paths for ion transport between opposing faces of the membrane. Severe mechanical stresses created by the swelling of the resin particles ultimately cause failure of the membrane. Mechanical problems also arise in homogeneous membranes, wherein woven substrates or skrim are used to enhance toughness, when adhesion is lost between the resin and the binder. As well, serviceability is determined by the chemical nature of the ion exchange sites. The chemical stability of cation exchange materials is much greater than for anion exchange resins, which contain quaternary ammonium groups. Cation membranes also generally provide good service at temperatures considerably above those allowed for anion membranes. In all respects, except ion exchange capacity in some cases, the homogeneous materials are superior, as are cation exchangers over their anion counterparts. By far the superior membranes are those based on the copolymers of tetrafluoroethylene and sulfonated or carboxylated perfluorovinyl ethers.

Initially aimed at the application of ion exchange separators in electrodialyzers for desalination, homogeneous membranes of the type derived by sulfonating or quaternizing cast films of basically styrene/divinylbenzene copolymers proved suitable. In Japan, in addition to desalination, table salt is produced from seawater by electrodialytic concentration. In such applications, as well as the demineralization of whey in cheese-making, both the media and the temperatures are mild with respect to chemical degradation of the membranes. Electrolytic cell applications on the other hand most often involve elevated temperatures and chemical species which destroy these membranes. Except for sulfonated styrene-divinylbenzene membranes developed for the electrohydrodimerization of acrylonitrile to adiponitrile (9, 29), no other significant applications of ion exchange membranes in electrosynthesis were made until the perfluorinated ionomer membranes became commercially available in the early 1970's.

Since that time a tremendous amount of work has been done and still continues on synthesis, applications, and understanding the basic physical chemistry of these revolutionary cation exchange membranes (7, 9, 17). For example, in the last two months newly announced conversions of existing chlor-alkali plants to electrolytic cells employing these membranes has increased from 4 to 6% of total world capacity, which is 182 billion pounds of products per year. The conversion of this entire industry over the coming years will cost about 4.5 billion 1984 dollars and provide about a 200 million dollar per year replacement market for the manufacturers of these membranes. The incentive is to save up to a third of the combined electrical and steam energy now used. This is equivalent to saving one trillion BTU per day when all plants are converted. The basis for these savings is the threefold contribution of the membrane in allowing more sophisticated cell designs to lower cell electrical energy consumption, limiting water transport to provide higher caustic concentrations in the cell effluent liquor, which saves steam in evaporation, and finally by limiting the concentrations of unwanted impurities such as NaCl in the product caustic to less than 100 ppm.

PROPERTIES AND APPLICATIONS OF ION EXCHANGE MEMBRANES

Ion exchange operates in the gel phase. The existence of hydrophilic ion exchange sites on the polymer backbone encourages the sorption of water and the swelling of the membrane. Without this wettability and formation of a gel phase there would be no ion exchange. Uncontrolled, however, there can be undesirable transport of water and co-ions across the membrane, leading to poor current efficiency. Co-ions bear the same charge as the exchange sites. The hydration of the membrane is highly influenced by the equivalent weight of the membrane, which is defined as the weight of polymer which will neutralize one equivalent of base or acid. The higher the equivalent weight, the lower the hydration. The transport of water in the membrane is also influenced by the nature of the counter-ion, i.e. the ion which is counter in charge to that of the ion exchange site, e.g. sodium ion in a sulfonic acid membrane. The electro-osmotic transport of water decreases with increasing concentration and valence of the counter-ion, and is usually greater for cationic as compared with anionic membranes. Further, it is directly proportional to the voltage gradient, and also influenced by the current density or rate of charge transport per unit of area. Overall the electro-osmotic transport of water is generally less than 30 gram moles per Faraday.

Equilibria between ion exchange membranes and electrolyte solutions are difficult to model. The classic thermodynamic model is the Donnan expression relating osmotic pressure, p, partial molar volume of a strong electrolyte, V, temperature, T, and the activity coefficient ratios of the anion and cation, respectively, in the solution, a, and in the membrane, a':

$$RT\ln\left[\left(\frac{a}{a'}\right)_A\left(\frac{a}{a'}\right)_B\right] = pV_{AB}$$

where AB is the electrolyte of cation A and anion B(18). However, use of the Donnan equation for other than very dilute solutions is made especially difficult by lack of knowledge either of the activity coefficients, particularly in the membrane, or of the osmotic pressure.

Strong efforts (19–24) have been made with some success toward predicting the performance of perfluorinated ion exchange membranes. Considering Eisenberg's (19) ion-cluster model, Fujimura (25) found that the size of the clusters is smaller for carboxylate than for sulfonate membranes, and decreases with both increasing temperature and decreasing equivalent weight, while increasing with cation size. Based on X-ray work Gierke (26) reported that the diameter of clusters in for example, a 1200 equivalent weight sodium sulfonate membrane is about 40 Angstroms. The clusters are connected by 10 Angstrom wide tunnels, and the individual clusters are spaced about 50 Angstroms apart. This model allows calculations of current efficiencies which are consistent with experimental observations.

Kimoto (24) used an empirical method to relate membrane hydration and Donnan equilibria in perfluoro membranes based on measurement of water content of equilibrated resins to determine the total fixed ion concentration in the resin, defined as 1000/(EW)(W), where EW is the equivalent weight and W is the grams of water/gram of "dry" resin. He then calculates osmotic pressure using an empirical formula relating EW, the molecular weight of the polymer in the hydrogen form, and, unfortunately, the activity of the water in very concentrated solutions. This can only be estimated (27) and further requires the Gluekauf (28) proportionality which relates osmotic pressure to the activity of the water raised to somewhere between the first and third power. Such attempts to define the parameters which constitute membrane performance are laudable, but have not yet allowed sufficient predictability, though correlation of the behaviors of *existing* membranes have been made.

Grot *et al.*, reported (20) that water absorption is a function of equivalent weight, salt form (i.e. the cation), and the previous history of the membrane. They expressed the weight % water take up in terms of two constants, one, A, dependent on the cation, and the other, B, dependent on the history of pretreatment of the membrane; also the water take up is an exponential function of the reciprocal of the equivalent weight:

$$\text{WT.\% } H_2O = 1.2AB \exp(3980/EW)$$

For the ions of K, Na and H, A is 0.57, 0.79, and 1, respectively. B depends on pretreatment but is 1 for the pre-boiled membrane. Since water absorption varies exponentially with increasing numbers of exchange sites, the higher the equivalent weight, the more effective the Donnan exclusion because of the decreased water absorption. Pretreatment by boiling in water, for example, was found (30) to lead to increased water absorption under normal cell use, whereas heating the dry membrane in a vaccum led to decreased water take up under normal cell use. These effects drastically influence current efficiency.

Seko (31) shows the effect of equivalent weight, acid strength of the exchange site (carboxylate or sulfonate), and weight % NaOH in the electrolyte on the absorption of water. The weak acid exchanger absorbs much less water than the strong acid exchanger, and both absorb less water as the concentration of NaOH in the electrolyte increases.

Ezzell and Carl (32) found that water absorption is more a function of the polymer molecular structure than anticipated, in the sense that channel size is determined by the extent of hydration per exchange site, as described by Hora and Maloney (33). Contrary to conventional wisdom these sulfonate membranes show very low hydration as compared with other sulfonic acid membranes. The latter contain long pendant chains separating the polymer backbone from the functional group, and hydration is so extensive that high equivalent weights are necessary to ensure Donnan exclusion. Another important expression of water absorption with respect to membrane functionality is based on both equivalent weight and the amount of hydration per functional group. Ezell and Carl (32) define this as the Hydration Product, that is the product of moles of H_2O per mole of exchange sites times the equivalent weight. On this basis it is interesting to note that long pendant group polymers have the same hydration product of around 22,000 at equivalent weights of essentially double those having smaller channel sizes.

To achieve good conductivity and high current efficiency in the production of highly concentrated caustic solutions requires tailoring the membrane to a given set of cell conditions such as current density, salt concentration, caustic concentration and required overall energy efficiency (34). This required that all earlier perfluorinated membranes be modified or abandoned. Bilayer membranes (35, 36) containing thin layers of carboxylic acid exchange sites were formed on, or bonded to sulfonic acid layers. Another approach (37) is to synthesize an entirely carboxylic membrane, but with differing equivalent weights on anode and cathode sides as was done earlier with all-sulfonic acid membranes in order to improve performance. The spate of activity in this highly competitive and potentially lucrative area is illustrated by recent U.S. Patent activity (32, and 38–54). The variations on the theme are many, and the applications of these fascinating materials will revolutionize more than the Chlor-Alkali industry as new applications are found.

REFERENCES

1. W. Juda and W.A. McRae, J. Am. Chem. Soc. *72* 1044 (1950)
2. W. Juda and W.A. McRae, U.S. Patent 2,636,851 (1953)
3. C. Hampel, Encyclopedia of Electrochemistry, Reinhold, New York (1964)
4. N. Lakshminarayanaiah, Chem. Rev. *65* 494 (1965)
5. R. Lacey and S. Loeb, Industrial Processing with Membranes, Wiley-Interscience, New York (1972)
6. H. Friedlander and R. Rickles, Anal-Chem. *37* 27A (1966)
7. D. Flett, ed. Ion Exchange Membranes, Ellis Horwood Ltd., Chichester (1983)
8. W. Juda et al., U.S. Patent 2,636,852 (1953)
9. M. Seko et al., Chap. 12 in Ref. (7)
10. J. Greatorex et al., On Site Generation of Hypochlorite (presented at the Water Pollution Control Federation meeting in San Francisco, Oct. 1971).
11. M. Minagawa and T. Ueda, Denki Kagaku *36* 589 (1968)
12. M. Yoshida et al., U.S. 3,654,104 (1972)
13. J. Leddy, J. Chem. Ed. *57* 640 (1980)
14. J. Leddy, Chap. 6 Handbook of Industrial Chemistry 8th. Ed., J. Kent ed., Van Nostrand, New York (1983)

15. D. Connolly and W. Gresham, U.S. 3,282,875 (1966)
16. W. Grot, U.S. 3,692,569 (1972)
17. A. Eisenberg and H. Yeager, ed. Perfluorinated Ionomer Membranes, Am. Chem. Soc. Symposium, Series 180, (1982)
18. F. Helfferich, Ion Exchange, McGraw Hill, New York (1962)
19. R. Yeo and A. Eisenberg, Polymer Preprints *16* 104 (1975)
20. W. Grot, G. Munn and P. Walmsley, Paper 154, Electrochem. Soc. Meeting, Houston (1972)
21. M. Seko, Paper 417, Electrochem. Soc. Meeting, Minneapolis (1981)
22. K. Mauritz, C. Hora, and A. Hopfinger, Chap. 8 in Ions in Polymers, ed. A. Eisenberg, ACS Advances in Chemistry Series, *187* (1980)
23. A. Hopfinger et al., Paper 439, Electrochem Soc. Meeting, Atlanta (1977)
24. K. Kimoto, J. Electrochem. Soc., *130* 334 (1983)
25. M. Fujimura et al., Macromolecules *15* 136 (1982)
26. T. Gierke, Electrochem Soc. Meeting, Atlanta (1977)
27. R. Robinson and R. Stokes, Electrolyte Solutions, Butterworths, London (1959)
28. E. Gluekauf, Proc. R. Soc. London, Ser. A, *214* 344 (1952)
29. J. Prescott, Chem. Eng. 72 (Nov. 8, 1965)
30. G. Munn, Electrochem Soc. Meeting, Atlanta (1977)
31. M. Seko et al., pg. 365, Ref. 17
32. B. Ezzell and W. Carl, U.S. Patent 4,417,969 (1983)
33. C. Hora and D. Maloney, Presented at the Electrochem. Soc. Meeting, Atlanta (1977)
34. C. Jackson, ed. Modern Chlor-Alkali Technology, Vol, 3, Society of the Chem. Ind., Ellis Horwood Publishers, Chichester (1983)
35. S. Ibrahim et al., pg. 53, Ref. (35)
36. M. Seko et al., pg. 76, Ref. (35)
37. M. Nagamura et al., pg. 61, Ref. (35)
38. M. Seko, U.S. 4,178,218 (1979)
39. T. Copeland and P. Resnick, U.S. 4,360,601 (1982)
40. D. Maloney, U.S. 4,349,422 (1982)
41. J. Solenberger and M. Withers, U.S. 4,341,605 (1982)
42. M. Withers, U.S. 4,324,606 (1982)
43. M. Seko, U.S. 4,357,218 (1982)
44. K. Kimoto et al., U.S. 4,332,665 (1982)
45. R. Silva and R. Smith, U.S. 4,401,711 (1983)
46. M. Withers, U.S. 4,399,183 (1983)
47. W. Grot, U.S. 4,415,679 (1983)
48. P. Resnick, U.S. 4,414,159 (1983)
49. K. Kimoto, U.S. 4,414,338 (1983)
50. M. Hamada et al., U.S. 4,425,199 (1984)
51. T. Seita et al., U.S. 4,431,504 (1984)
52. T. Bissot et al., U.S. 4,437,951 (1984)
53. R. Smith and M. Withers, U.S. 4,437,952 (1984)
54. M. Yoshida et al., U.S. 4,426,271 (1984)

DISCUSSION

ANSWER (Question inaudible) First of all the hydration product, of two very important variables the equivalent weight. And also the amount of water per functional group. You measure the amount of water uptake and relate it to moles of water per mole of functional group.

QUESTION In the membrane you have free water and water hydration and if you have different ions the volume of water in the membrane will change up to several hundred percent.

ANSWER In this case, these are in sodium form. The idea of the exercise is to show the difference, especially at a given equivalent weight. But, it's easy to measure the amount of water and it's easy to measure the equivalent weight and it is easy then to relate the absorbed water to the number of functional groups.

QUESTION It's a construct?

ANSWER I guess I'm not sure exactly what your question is.

QUESTION It's very easy to measure the amount of water in the polymer and to divide by milliequivalent per gram?

ANSWER That's all this is, I know what his question is now. Do we take account of the other water? The answer is no. So go ahead with your point.

QUESTION I don't want to make a point, I just wanted to know whether you calculate this or not.

ANSWER Oh, I can answer that. I didn't understand your question at first. There's no question that the water measured is absorbed water. I assume, "dry membrane", but not pretreated dry. You can alter the results tremendously.

QUESTION The patent up to that time taught the only known way of affecting the hydration per functional group at a given equivalent weight is by changing a functional group. The chemistry that was available for making these fluoro polymers up to this time allowed only the longer side chain, because in the process of making the monomer if you had the shorter side chain you got cyclization in a previous step that prevented making the monomer. Ezil and Carl found a way around that and were able to make the short size change which was one ether linkage two carbons before the sulfonic. In doing so, at a given equivalent weight, you translate a huge portion of the molecules of the polymer from the side chain into the backbone. Effectively, you created a greater resistance to the osmotic pressure at a given equivalent weight. You were effectively squeezing out the free water present in the system. The net result was a higher current efficiency for the same functional group, same equivalent weight, because the amount of free water was reduced?

ANSWER There is no question, because we're talking about smaller diameters. Were these essentially normalized to constant conversion? People tend to state their fluxes and rejection based on 50–70% conversion. With spirals you're always starting from 15–20% conversion. Will they normalize? No, I really can't speak on this. I borrowed this from Kurihara of Toray Industries. He put everything on a logarithmic scale and I put it on an arithmetic scale which I think is perhaps more instructive. I don't know whether he did that kind of normalization or not.

QUESTION I have an uncertainty about the design of a structure such that you can open it up and increase the free volume. Is that what you are mentioning?

ANSWER In glassy polymers I hesitate to use the words "free volume" as much as the word sometimes used "unrelaxed volume". In rubbers it's much clearer what is meant by "free volume". In glassy materials which are nonequilibrium in their nature, people distinguish between "free volume" and "unrelaxed volume". I would say there is more interchain or intersegmental spacing between the polymer chains and I think the most effective materials are those that have been heat treated down to the point where that's not going to continue to change over a reasonable time scale. So in that context, I would mean "free volume".

QUESTION Is there any physical method by which we can quantify the space?

ANSWER What I believe is the most effective way to do this is literally to accept the gas molecule as the most sensitive probe of molecular structure that's around. Decide upon what is the effective size of that gas molecule and the best way to know is to look at these kinetic diameters from zeolites where we can talk with some certainty about "that molecule can get into that space", and then use that — it's still a correlating variable, but I think it's a powerful one. Then start producing "spectra" of the size frequency of these gaps, which is what the diffusion coefficient is to a first approximation. The reason I said it's a little bit simple minded is of course there can be differences in jump length, depending upon what the molecule is, but those tend to be 15–20% differences. The real difference that controls the diffusion coefficient is the frequency of gaps.

My conclusion is that if the measurements are made well, these diffusion coefficients are the most sensitive. We're talking about angstrom and subangstrom characterization.

Membranes for Desalination

R. J. PETERSEN

FilmTec Corporation, Minneapolis, Minnesota

A variety of reverse osmosis membranes are now commercially available for water desalination. Examples include both cellulose acetate and aromatic polyamide membranes in sheet and hollow fiber form, and various thin-film-composite membranes which contain either polyamides or sulfonated polymer resins as the active desalination layer. The performance advantages and disadvantages of these various membranes will be reviewed. Although numerous reverse osmosis membranes have been described in the technical literature, the emphasis in this presentation will be on membranes that have reached commercial status, or are believed to be approaching commercialization. New directions in reverse osmosis membrane will be assessed, including the appearance of "loose" membranes for industrial separations. Also, the potential of microporous membranes for membrane distillation as a new route to potable water from seawater will be covered.

INTRODUCTION

The guiding theme in this paper is the practical aspects of membranes for desalination. Specifically, the origin, types, and performance characteristics of current, commercially available reverse osmosis desalination membranes are reviewed. Also, indications of recent directions in reverse osmosis membranes are discussed. Membrane distillation is briefly examined as well.

Two quite recent reviews of membrane processes are worth noting. One is a brief practical overview of membrane separation processes, including reverse osmosis, by Applegate. (1) The second is the newly published book "Synthetic Membrane Processes" edited by Belfort. (2)

GENERAL CHRONOLOGY OF REVERSE OSMOSIS MEMBRANE DEVELOPMENT

Figure 1 contains a chart showing the chronological unfolding of key developments in the field of reverse osmosis desalination membranes, as perceived by this author. Reid and Breton (3) first demonstrated the

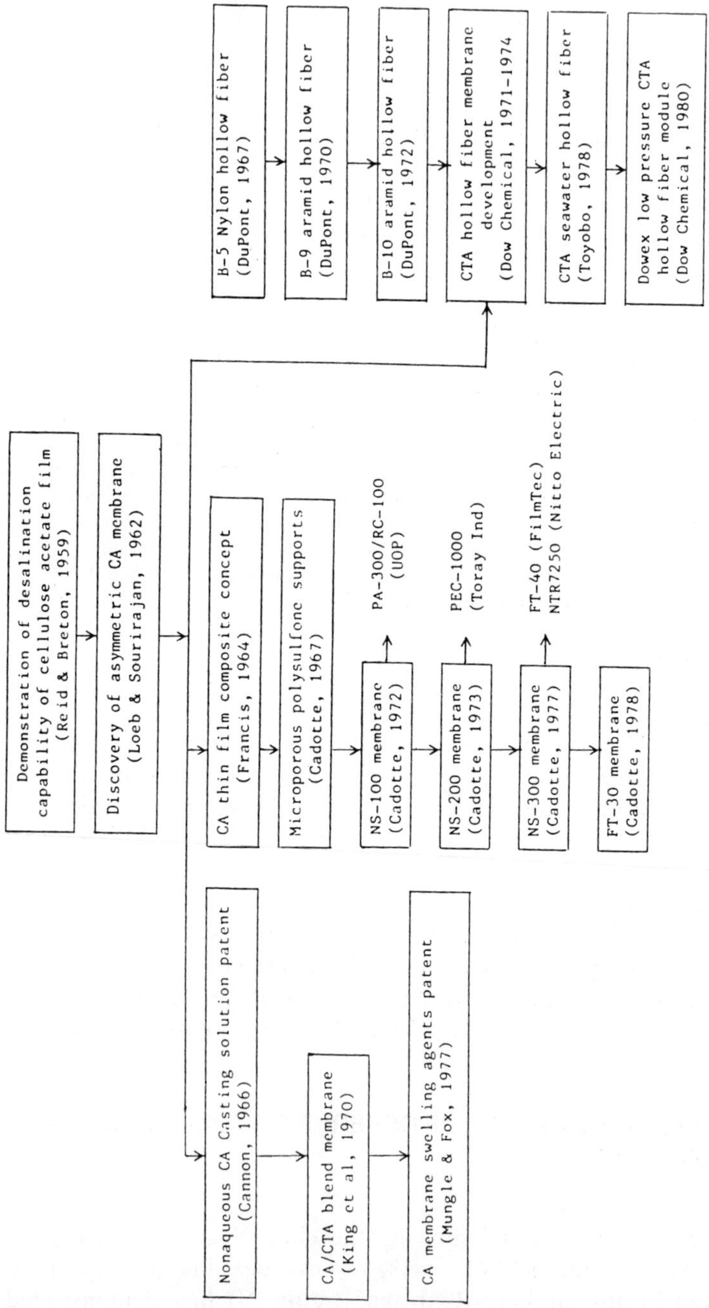

FIGURE 1 Chronological Chart of Reverse Osmosis Membrane Technology Highlighting Key Developments in Membrane Commercialization.

ability of cellulose acetate films to reject salt while allowing water permeation under high transmembrane pressure. Soon thereafter, Loeb and Sourirajan (4) discovered the technique of making asymmetric cellulose acetate membranes, and the practical utility of these membranes in producing high water fluxes while retaining salt rejection characteristics. The key to this development was the fabrication of a film comprised of an ultrathin, dense, semipermeable surface skin supported on a thick, microporous substructure. The surface skin provided the desalinating characteristics, and the sublayer provided thickness and handling strength while contributing very low impedance to permeate flow.

Various improvements have been made in the formulation and casting of asymmetric cellulose acetate sheet membrane since its initial discovery. Notable achievements include the development of nonaqueous casting dopes for cellulose acetate membranes, (5) the development of the blend membrane, (6) and the art of using swelling agents in conjunction with the casting of this membrane. (7) Meanwhile, efforts to extract greater flux efficiencies from cellulose acetate membranes led to two other developments: thin film composite membranes and hollow fiber membranes.

Francis (8) developed the approach of separately fabricating ultrathin cellulose acetate films and microporous cellulose ester sheets, then laminating these together to form thin-skinned membranes. In theory, each layer in the thin film composite membrane could be separately optimized for performance characteristics. A potentially commercial version of this approach was later developed by Lonsdale and coworkers, (9) but was overtaken by rapid improvements in the flux characteristics of asymmetric cellulose acetate membranes.

An important development in the thin film composite approach was the discovery of microporous polysulfone support films in 1967 by Cadotte. (10) Not only did this result in twice the flux for ultrathin cellulose acetate composite membranes, but it led subsequently to the invention of four important non-cellulosic thin film composite membranes by Cadotte: NS-100, (11) NS-200, (12) NS-300 (13) and FT-30 (14) Each of these discoveries is either embodied in a commercial reverse osmosis membrane, or has led to a modified, closely related commercial membrane. Microporous polysulfone itself has achieved widespread use as an ultrafiltration membrane.

Hollow fiber technology represented an alternate route to the development of commercial reverse osmosis membranes for desalination. The initial entry in this field was the Aramid (aromatic polyamide) hollow fiber membrane of E.I. duPont de Nemours & Company. (15) The DuPont hollow fiber membrane technology apparently had been originally developed with gas separations in mind as the intended market application. However, the discovery by Richter and Hoehn (16) of the utility of Aramid membranes for desalination led to the development of the B-9 and B-10 Permasep hollow fiber permeators for brackish water and seawater applications respectively. (Actually, this development was preceded by the B-5 aliphatic polyamide hollow fiber permeator, which, however, was found in field applications to be damaged by dissolved oxygen.)

The Aramid hollow fiber membranes were characterized by quite low flux in comparison with asymmetric and thin film composite sheet membranes. However, the ability to package an enormous amount of hollow fiber surface area into a small volumetric space resulted in favorable device productivity and efficiency. It should be noted that these hollow fiber membranes are indeed classic asymmetric membranes similar in cross-section to the Loeb-Sourirajan membranes. They contain a dense, semipermeable, outer skin and a porous structure throughout the remainder of the fiber wall.

Similar work was carried out by the Dow Chemical Company on cellulose acetate and cellulose triacetate hollow fiber compositions. (17, 18) Dow has since developed a commercial hollow fiber system for low pressure reverse osmosis purification of brackish water sources. Following the general approach of the early Dow work, Toybo Company has succeeded in developing in high pressure seawater desalination hollow fiber of cellulose triacetate, as well as a higher flux, moderate pressure version for brackish water treatment. (19)

All of these membrane types are related in that all of them employ a thin, semipermeable, desalinizing layer 500 to 2500 angstroms thick. All employ a porous substructure as a supporting layer for this thin, permselective membrane layer. This type of structure is illustrated by the photomicrograph in Figure 2. In asymmetric membranes (sheet and hollow fiber), both layers of the membrane are of the same composition and integrally bonded. In thin film composite membranes, the two layers are of differing composition and are fabricated in a two-step process rather than in a single step.

CELLULOSE ACETATE MEMBRANES

Asymmetric cellulose acetate membranes were of course, the first reverse osmosis membranes to be developed commercially. These membranes still occupy about half the installed capacity in the world. There are several major manufacturers of this membrane type. In sheet form, this membrane is relatively easy to fabricate by today's state of the art.

FIGURE 2 Photomicrograph of a Reverse Osmosis Membrane Showing the Smooth Dense Surface Skin and a Cross-section of the Microporous Sublayer.

The chemical structure of cellulose acetate is shown below:

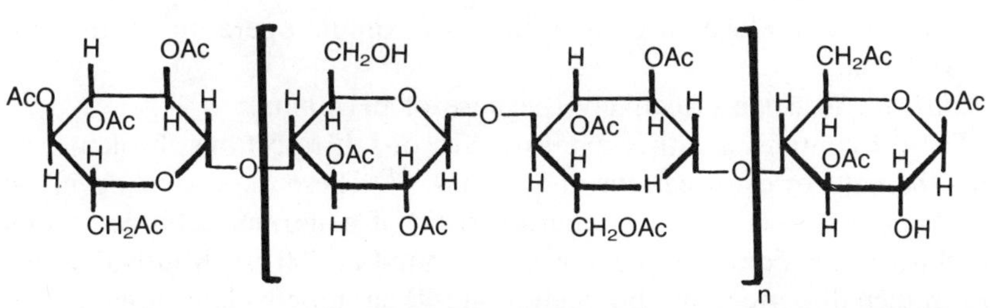

A fully acetylated cellulose polymer will contain three acetate groups per saccharide ring, hence the name cellulose triacetate. Initial reverse osmosis membranes were based on "secondary cellulose acetate", which contained an average of about 2.7 acetate groups per ring. The residual free hydroxyl content contributed to higher water flux. Cellulose triacetate membranes are more chemically stable and provide higher salt rejections, but at a very significant loss in flux. The problem of low flux has been overcome, of course, by use of hollow fiber technology. Blend membranes contain both cellulose triacetate and secondary cellulose acetate. One can obtain significantly better flux and salt rejection characteristics with a properly formulated blend membrane than is possible with a homopolymer cellulose acetate grade with the same acetate content as the average in the blend membrane.

The most notable manufacturers of cellulose acetate reverse osmosis membranes include:

a) spiral wound elements: UOP, Toray Industries, Hydranautics, Desalination Systems, Osmonics, Envirogenics;
b) hollow fiber modules: Dow Chemical, Toyobo;
c) tubular modules: Paterson Candy International (PCI), Nitto Electric Industrial;
d) plate and frame modules: De Danske Sukkerfabrikker (DDS).

This is by no means an exhaustive list. Many smaller manufacturers of cellulose acetate membrane and elements exist.

Cellulose acetate membranes have various advantages, including:

a) being inexpensive relative to competing membranes;
b) more easily fabricated than other membrane types;
c) well characterized (20 years of operating data, innumerable research studies published); and
d) resistant to chlorine (in chlorinated feedwater).

Reverse osmosis membranes of cellulose acetate suffer from several disadvantages, however, including:

a) generally lower solute rejections than other membranes;
b) easy susceptibility to hydrolysis (operating limits of pH 3–8);
c) biodegradability;
d) sensitivity to elevated temperatures (maximum operating temperature 35°C); and
e) difficult to clean effectively (because of pH limits).

Table I contains a rather extensive list of salt rejections, broken down into the categories of cations and anions.[20] These data were developed by Osmonics, Inc., and are characteristic of a normal cellulose acetate brackish water desalting membrane operated at 400 psi. Multivalent ions are rejected better by the membrane than monovalent ions. Where different ion valences are combined, salt passage is approximately controlled by the ion with the highest valence.

TABLE I

Typical Rejections of Dissolved Salts by Spiral Wound Cellulose Acetate Membranes

Cations		Percent Rejection	Percent Passage (Average)	Anions		Percent Rejection	Percent Passage (Average)
Sodium	Na^{+}	94–96	5	Chloride	Cl^{-}	94–95	5
Calcium	Ca^{++}	96–98	3	Bicarbonate	HCO_3^{-}	95–96	4
Magnesium	Mg^{++}	96–98	3	Sulfate	SO_4^{--}	>99	1
Potassium	K^{+}	94–96	5	Nitrate	NO_3^{-}	85–95	10
Iron	Fe^{++}	98–99	2	Fluoride	F^{-}	94–96	5
Manganese	Mn^{++}	98–99	2	Silicate	SiO_2^{--}	80–95	10
Aluminum	Al^{+++}	>99	1	Phosphate	PO_4^{---}	>99	1
Ammonium	NH_4^{+}	88–95	8	Bromide	Br^{-}	94–96	5
Copper	Cu^{++}	98–99	1	Borate	$B_4O_7^{--}$	35–70	—
Nickel	Ni^{++}	98–99	1	Chromate	CrO_4^{--}	90–98	6
Zinc	Zn^{++}	98–99	1	Cyanide	CN^{-}	90–95	—
Strontium	Sr^{++}	96–99	3	Sulfite	SO_3^{--}	98–99	1
Cadmium	Cd^{++}	96–98	3	Thiosulfate	$S_2O_3^{--}$	>99	1
Silver	Ag^{+}	94–96	5	Ferrocyanide	$Fe(CN)_6^{---}$	>99	1
Mercury	Hg^{++}	96–98	3				

The solute rejection characteristics of cellulose acetate membranes can be tailored downward as well as upward. Table II contains data for high, medium and low retention membranes as manufactured by the Danish sugar company, DDS, for use in its plate and frame systems.

Hydrolysis of the surface skin in cellulose acetate membranes results in a loss of retentivity. This is one of the major drawbacks affecting its use. Figure 3 shows hydrolysis rate data gathered by Vos and co-workers.[(21)] Hydrolysis rates showed a minimum at pH 4.5.

Superimposed on this graph are the data of Spatz and Friedlander showing loss of salt rejection as a function of pH. These workers measured initial salt passage rates for a group of cellulose acetate membrane specimens, then immersed these specimens in water adjusted to various pH values. Three week exposures at 60°C were used (estimated to correspond to about 8 to 10 months exposure at 25°C), followed by retesting of the membranes. The ratio of final salt passage to initial salt passage, plotted at the bottom of Figure 3, begins to rise for membrane exposures beyond the range of pH 3–7. Spatz and Friedlander estimated membrane life to be three years for pH 3–7.2, and one year for pH 2.5–3 and pH 7.2–8.5. Many naturally occurring waters have pH values in the range of 7.5 to 9, so that feedwater acidification is required.

ARAMID HOLLOW FIBER MEMBRANES

The Permasep hollow fiber permeators introduced by DuPont in 1970 represented the first commercial membrane alternative that overcame some of the drawbacks of cellulose acetate membranes. These Aramid

TABLE II

Permeability data for low molecular weight organic compounds through Various Grades of Cellulose Acetate Membranes

	% permeability*					
		phenol	ethanol	urea	1.5% L-lactic acid	
Membrane type	NaCl Solution	0.003–0.10% solution	0.9–2.6% solution	0.1–2.0% solution	pH = 7	pH = 3
990	>15	~117	88–89	80	4	48–52
995	> 6	~107	86–88	70	3	34
999	> 2	~107	77–78	50	0.5–1.0	20

* Measured at 40 bar, 25°C, flow 8 1/min, DDS module, 20.
Source: De Danske Sukkerfabrikker product literature.

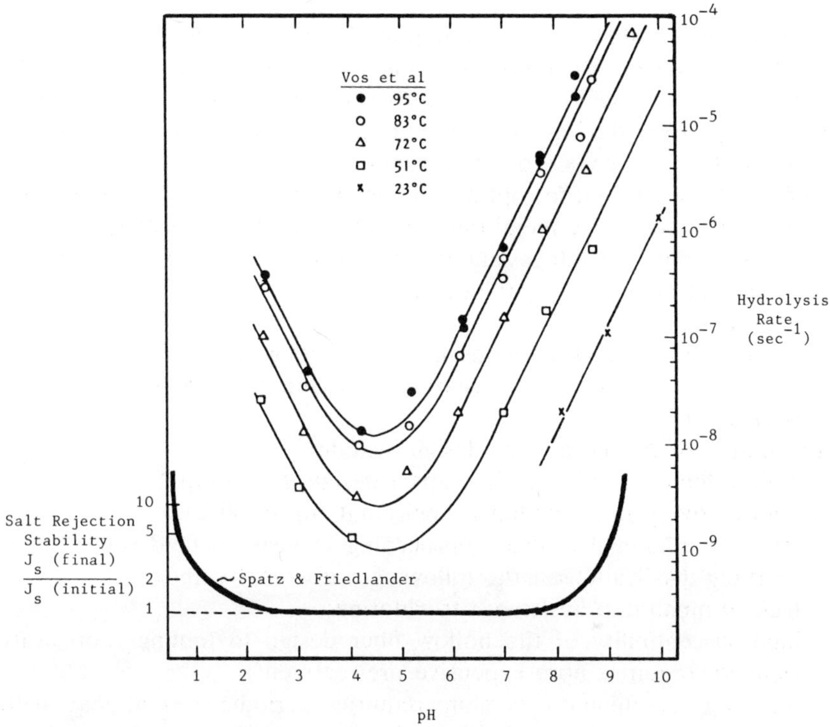

FIGURE 3 Cellulose Acetate Membrane Hydrolysis Rate and Salt Rejection Stability Versus pH.

hollow fiber membranes could be operated at a pH range of 4–11, were far more resistant to hydrolysis, and were not biodegradable.

The chemical structure of the Aramid polymer used in these hollow fiber membranes is believed to be approximately as follows:

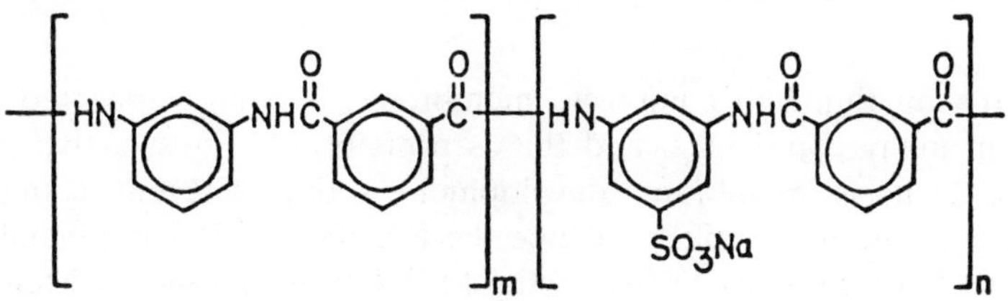

The Permasep hollow fiber modules are made with extruded hollow fibers having an outside diameter of 85 μm and an internal diameter of 42 μm. The fibers are distributed crosswise upon a spunbonded polyester web, wound upon a hollow mandrel, and potted at each end in epoxy resin. At one end, the face of the epoxy resin plug is machined away to expose the hollow fiber openings (called the "tube sheet"). Feedwater is introduced into the unit from the hollow core, and flows radially to the outside of the fiber bundle. The choice of fiber dimensions and element design have been carefully selected for optimized performance of the finished design, taking into account the membrane flux and bore flow resistance.

There are both advantages and disadvantages to this particular product.

a) broad pH range (4–11) tolerance;
b) nonbiodegradability;
c) higher solute rejections (reduced salt passage, good organic rejections);
d) first true "single stage" seawater desalination membrane (B-10 permeator).

Advantages of this permeator design include:

a) high volumetric efficiency (compact design, high output);
b) high recovery ratios within a single unit (up to 30%);
c) strong commercial record (approaching 15 years of field experience).

Concerning disadvantages, the following are worth noting:

a) lack of membrane resistance to chlorine;
b) high susceptibility of the hollow fiber design to fouling (not easily cleaned, requires more expensive pretreatment);
c) seawater desalination version requires periodic retreatment with tannic acid colloid to maintain high salt rejection;
d) sensitivity to temperature (maximum 35°C).

Aramid hollow fiber modules have taken over a large segment of the market, but are now beginning to meet strong competition from both cellulose acetate hollow fiber and thin film composite membrane spiral devices in the market place.

THIN FILM COMPOSITE MEMBRANES

Research on thin film composite membranes is being supported extensively in many companies, and this is probably the most active area in reverse osmosis membrane development today. Current commercial offerings in the marketplace include the PA-300 and RC-100 membranes by UOP Fluid Systems Division, the MSI-400 membrane by Membrane Systems (a sister company of Hydranautics), the PEC-1000 membrane by Toray Industries, the FT-30 and FT-40 membranes by FilmTec, and the NTR-7250 membrane by Nitto Electric Industrial Company.

The original NS-100 membrane, invented at North Star Research

Institute by Cadotte, (11) was a thin film composite membrane comprised of a microporous polysulfone base layer, a topcoating of polyethylenimine, and an interfacially formed skin of polyurea or polyamide produced by contact of toluene diisocyanate or isophthaloyl chloride with the polyethylenimine coating. The barrier layer is only about 200 angstroms thick. The remaining unreacted polyethylenimine is insolubilized by heat-curing at 110°C. Formation of this membrane is illustrated in Figure 4. The MSI-400 membrane is believed to be identical to NS-100 in composition. (23)

The PA-300 and RC-100 membranes of UOP differ from NS-100 specifically in the choice of a different polyamine called "polyepiamine." (24) This polyamine is apparently made by reaction of polyepichlorohydrin with ethylenediamine, and has the following nominal chemical structure:

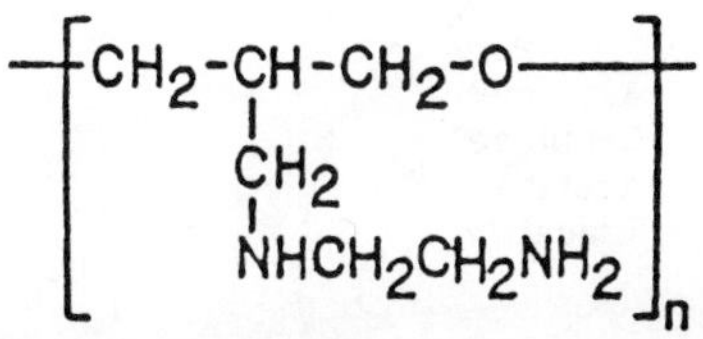

In PA-300, isophthaloyl chloride is used as a crosslinking agent, producing a polyamide barrier layer. In RC-100, toluene diisocyanate is believed to be the crosslinking agent, generating a polyurea barrier layer. Closely related versions of these membranes are believed to have been developed as well by Desalination Systems and Hydranautics.

Desalination membranes with good flux and high salt rejections are possible by this approach. In fact, a 3.2-million gallon per day plant was installed at Jedda, Saudi Arabia, by UOP using the PA-300 and RC-100 membranes in spiral element form. (25) A high flux, low pressure version has been developed for brackish water applications.

The advantages of this membrane include excellent pH resistance, high solute rejections, and good flux. However, the extreme thinness of the salt barrier layer has resulted in marginal performance, particularly in seawater applications. This occurs because the extremely thin salt barrier layer is susceptible to physical and chemical damage. Loss of salt rejection due to chlorine exposure, for example, is extremely rapid. These membranes are all weakly cationic, which makes them susceptible to fouling by ordinary anionic surfactants. This would limit their usefulness in potential wastewater applications.

The NS-200 type of membrane would be expected to be considerably more rugged than membranes of the NS-100 type. NS-200 is formed by impregnating microporous polysulfone sheeting with an aqueous solution containing furfuryl alcohol and sulfuric acid. The impregnated film is then

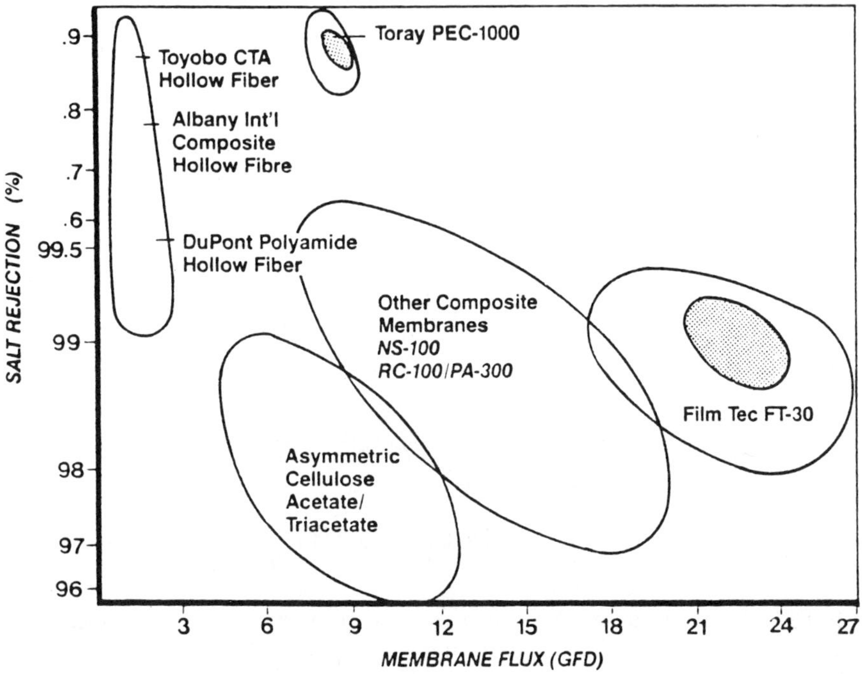

FIGURE 4 Comparison of Membrane Performance Levels in Seawater Desalination (3.5% Seawater, 800 psi, 25°C).

heated to 135°C. This results in the formation of a brown-colored sulfonated polyfurane resin. Because of the low viscosity of the coating solution and the low molecular weight of the chemical species contained therein, the resulting barrier layer is embedded in the surface rather than sitting on top of the surface. The thickness of the resin layer is about 500 angstroms. Consequently, it would be more resistant to physical abrasion.

The NS-200 membrane exhibited excellent salt rejections — typically 99.8–99.9 in synthetic seawater tests at 1500 psi. However, it was found to be unstable, and lost rejection with time during reverse osmosis operation. Efforts by several U.S. companies to make a viable commercial version of NS-200 ended in failure. The problem of its instability appears to have been solved by a Japanese company, Toray Industries. Beginning with isocyanurate chemicals as building blocks, Kurihara and co-workers eventually developed a hybrid membrane containing 1,3,5-tris(hydroxyethyl)isocyanurate, furfuryl alcohol and sulfuric acid. (26) Though never stated explicitly, this is believed to be the structure of the Toray PEC-1000 membrane.

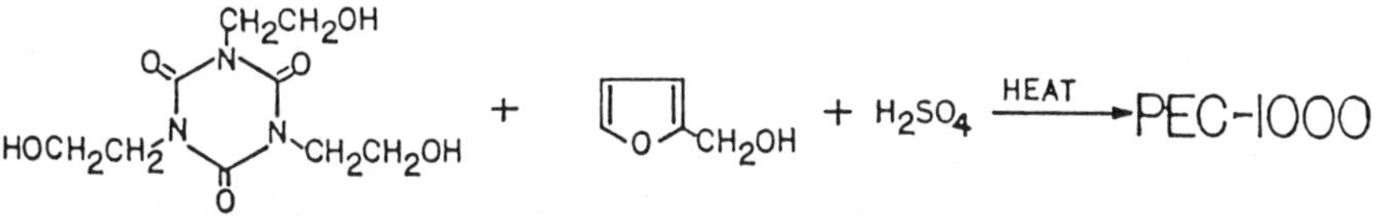

Key attributes of this reverse osmosis membrane include its excellent salt rejection and excellent retention of various low molecular weight solvents such as alcohols and ketones. (27) Thus, it is capable of single pass seawater desalination on even high salinity feedwaters such as Persian Gulf seawater. No other membrane matches it for organic rejections either.

The PEC-1000 membrane does suffer some disadvantages, however. It is not resistant to chlorine. Furthermore, it is susceptible to attack by dissolved oxygen in the feedwater. Either deoxygenation of the feedwater or continuous dosing of the feedwater with sodium bisulfite is required to maintain membrane integrity. Finally, its flux is low, such that its potential range of market applications is limited.

The FT-30 membrane is prepared by interfacial polymerization of metaphenylene-diamine with trimesoyl chloride to give a barrier layer of the following approximate composition:

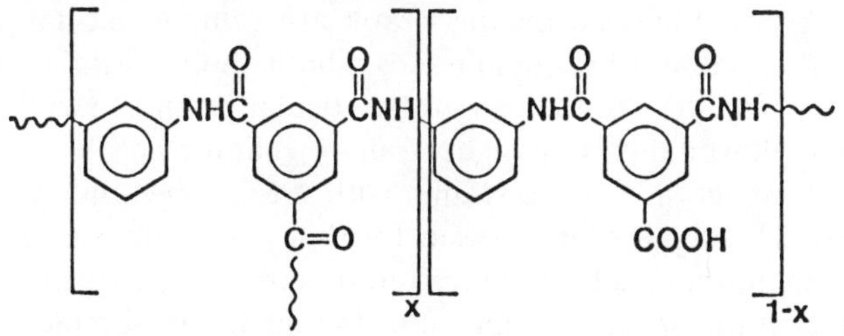

Thus, it is rather similar to the DuPont Aramid compositions, and its solute rejection characteristics are basically equivalent to those of the Aramid hollow fibers. (28) However, the extensive crosslinking in the FT-30 aromatic polyamide results in better chemical resistance, including some resistance to accidental chlorine exposure. Also, periodic treatment with tannic acid is not necessary to maintain salt rejection.

The barrier layer in FT-30 thin film composite membranes averages 2000 angstroms thick, versus 100–200 angstroms in the NS-100/PA/300/RC-100/MSI/400 series and 500 angstroms in the PEC-1000 membrane. This results in FT-30 being a considerably more rugged membrane, capable of harsher operating and cleaning environments. It can be continuously operated at pH 2–12. In the form of spiral elements, it can be operated at temperatures to 45°C. In plate and frame devices, operation at temperatures as high as 80°C is possible. Also, it has the highest flux of any known reverse osmosis membrane having equivalent salt rejection characteristics.

Advantages include, therefore, the following:

a) high flux (18–22 gfd at 225 psi, 24–28 gfd on seawater at 800 psi);
b) high rejections (96–98% on brackish water, 99+% on seawater);
c) broad pH capability (pH range 2–12);
d) higher operating temperature limits:
e) some minimal chlorine tolerance (200–1000 ppm-hrs);
f) ability to be effectively cleaned (because of broad pH tolerance);
g) maximum salt rejection at pH 8–8.5.

Some disadvantages of this membrane include:

a) lack of long term chlorine resistance;
b) increased sensitivity to flux loss by fouling;
c) lack of long term field experience.

COMPARISONS BETWEEN MEMBRANES

Figure 4 is a graph of salt rejection versus membrane flux for a variety of membrane types. Included on this graph are cellulose acetate, Aramid, and thin film composite membranes, both hollow fiber and sheet membranes. One can see that, in any particular group, there is a general trend toward lower fluxes at higher salt rejections, and vice versa. The highest salt rejections are possible with PEC-1000 and hollow fiber membranes. The highest fluxes occur for thin film composite membranes. Whichever membrane is to be preferred on cost and performance criteria will depend on the specific application, the quality of the incoming feedwater, and the desired quality of the permeate. No broad, encompassing recommendations can be made. Market trends currently favor, however, the hollow fiber devices and the newer thin film composite membrane spirals at the expense of cellulose acetate spirals and the NS-100 membrane thin film composite types.

Figure 5 compares organic rejections for several types of membranes. This figure, adopted from Kurihara and co-workers, shows the general range of organic solute rejections that one can expect from cellulose acetate (bottom line), PEC-1000 (top line) and FT-30 (basically representative also of DuPont Aramid) (middle line). The PEC-1000 membrane is obviously superior for organic rejections.

It must be noted, however, that the reverse osmosis process is not necessarily an economical and effective process for concentration of small

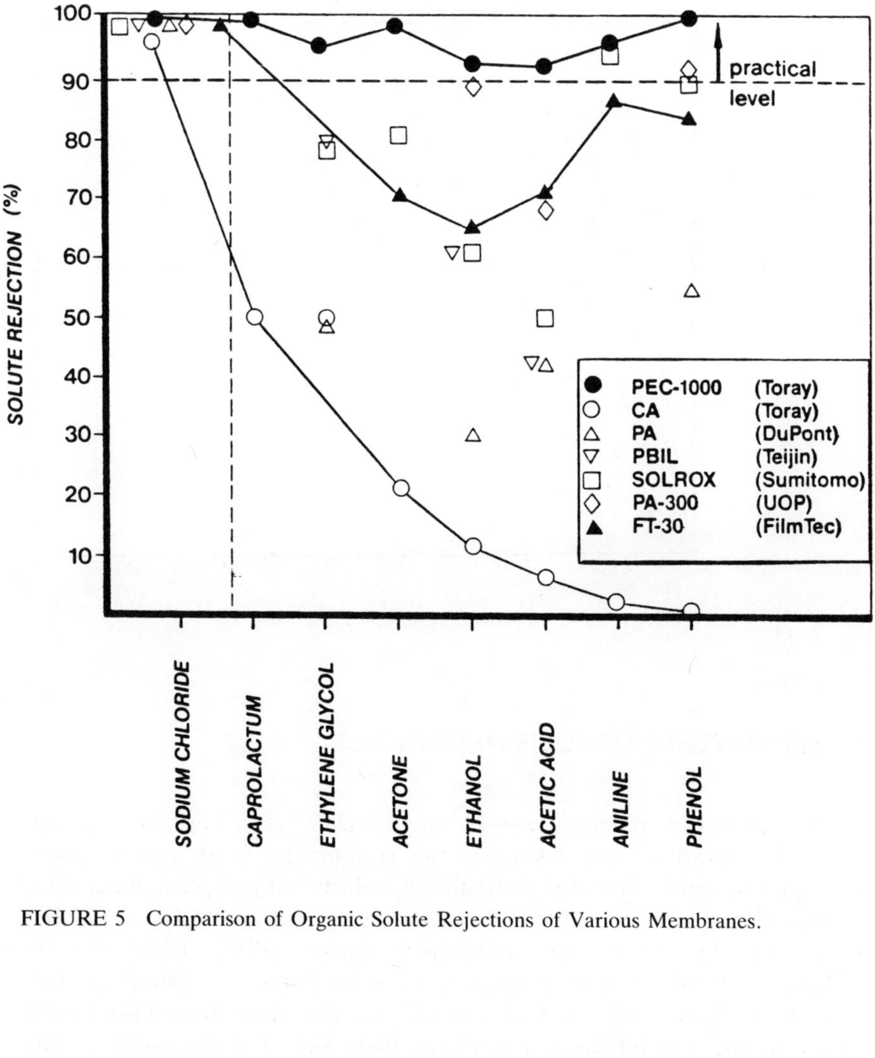

FIGURE 5 Comparison of Organic Solute Rejections of Various Membranes.

organic chemicals such as ethanol. Small organic chemicals create osmotic pressure in aqueous solutions, just as occurs with dissolved salts. At some point, probably in the neighborhood of 500 psi osmotic pressure, the reverse osmosis process becomes rather energy intensive because of the high pumping pressures required to overcome the osmotic pressure. Figure 6 illustrates the osmotic pressure effect of several solutes as a function of concentration. Ethanol, as an example, generates 500 psi osmotic pressure at about six percent concentration in water.

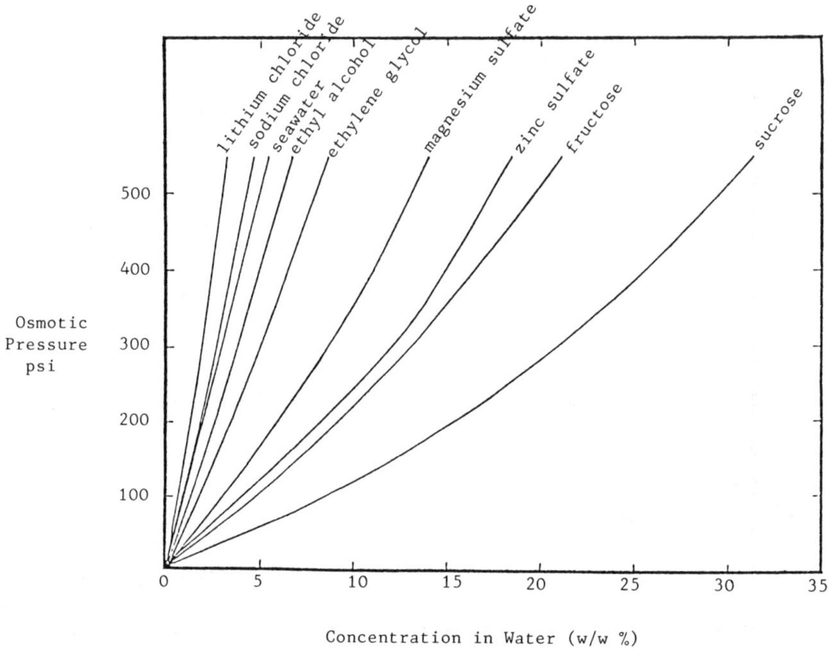

FIGURE 6

OTHER REVERSE OSMOSIS MEMBRANES

Numerous other reverse osmosis membranes have appeared in the scientific literature, and some on the commercial marketplace. Some examples include sulfonated polysulfone, sulfonated polyphenyleneoxide, asymmetric polypiperazineamides, the polybenzimidazolone (PBIL) membrane by Teijin, the polybenzimidazole (PBI) membrane by Celanese, and the Solrox 2000 membrane by Sumitomo Chemical. For one reason or another, these membranes are not at this time a significant factor in the marketplace. Therefore, they were not discussed in this paper. Their individual market status could, of course, change in the near or distant future.

SOME BASIC REVERSE OSMOSIS PERFORMANCE FACTORS

Some of the basic performance factors that one sees for all reverse osmosis membranes are plotted in the next five figures, using data from

FT-30 membrane tests for illustration. Water flux and salt rejection as the function of pressure are plotted in Figure 7. As operating pressure goes up, water flux also increases, following almost a direct relationship with pressure. The intercept of this line at the abscissa represents the osmotic pressure of the feed solution. This type of relationship will be true for all kinds of membranes, so long as one does not exceed either the ability of the membrane to resist compaction or the ability of the finished membrane device to handle high permeate flows. Salt rejection, on the other hand, follows a curvilinear path, reaching a salt rejection plateau when sufficient water flux is being forced through the semipermeable membrane. Reverse osmosis desalination membranes are not perfect barriers to dissolved salts in the feedstream. There is always an underlying salt transport through the barrier layer. This is overcome to a large extent by pushing water through at a faster rate than the salt can go through, since salt transport is not directly related to applied pressure. There is a limit on the high end, a plateau in which some salt flow is coupled with water flow through the membrane.

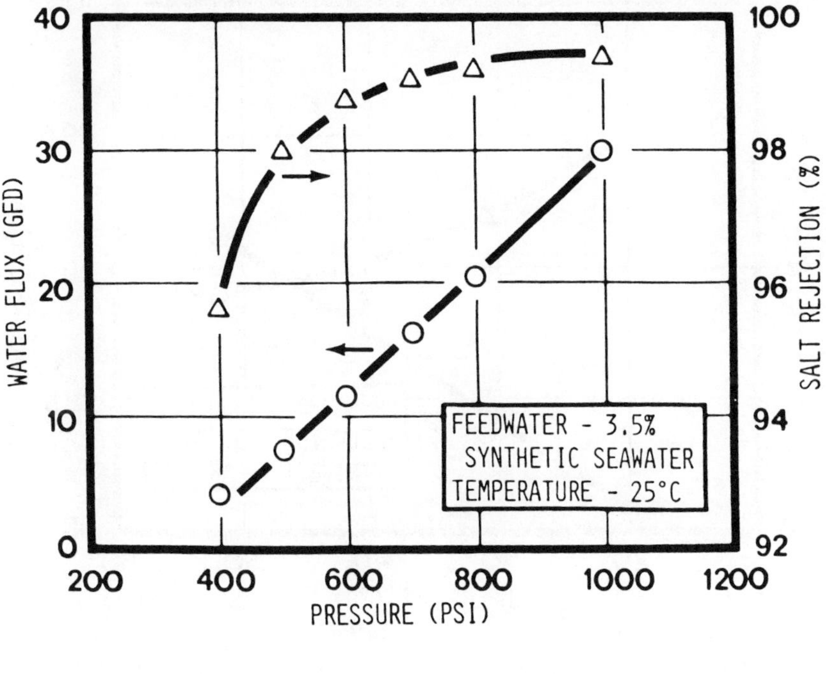

FIGURE 7 Effect of Feedwater Pressure on Flux and Salt Rejection of an FT-30 Thin Film Composite Membrane.

Figure 8 illustrates the fact that, as the temperature of the water is increased, membrane flux increases. This is due primarily to lower viscosity of the water, and higher diffusion rate through the membrane. It can also reflect some probable swelling of the membrane by water solvation. Both effects would occur as temperature goes up. A drop-off in salt rejection as a function of temperature also occurs. This is probably due both to the water swelling effect on the membrane structure, and to increased solubility of salt in the membrane. In the case of the FT-30 membrane, a temperature correction factor of 3 to 4 percent per degree Celsius is used to predict membrane flux.

Figure 9 shows the effect of feedwater concentration on flux and salt rejection. As the concentration of the feedwater increases, water flux will drop off. It is mainly due to the increasing osmotic pressure that one must subtract from the system operating pressure to get the net driving pressure for water transport. To a lesser extent, increased feedwater concentration also leads to a stronger concentration polarization effect. A

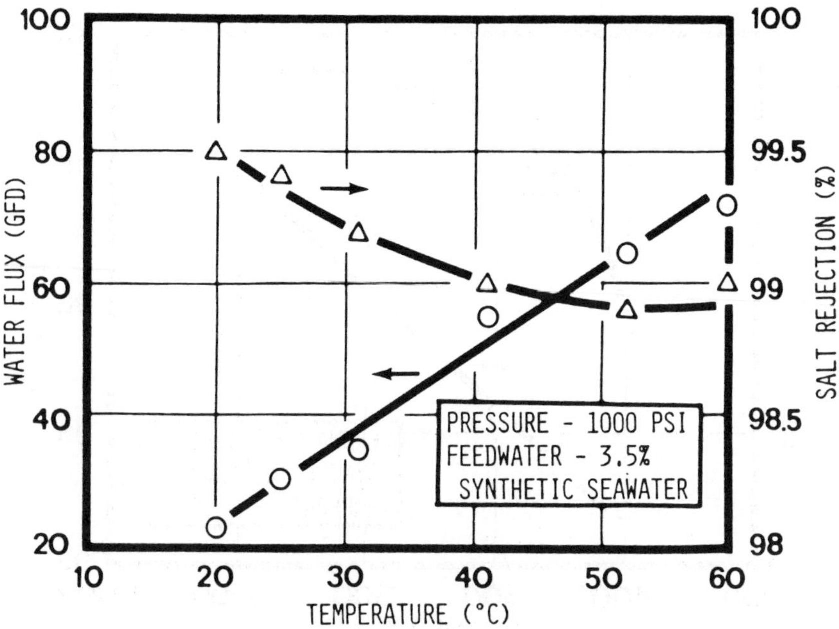

FIGURE 8 Effect of Feedwater Temperature on Flux and Salt Rejection of an FT-30 Thin Film Composite Membrane.

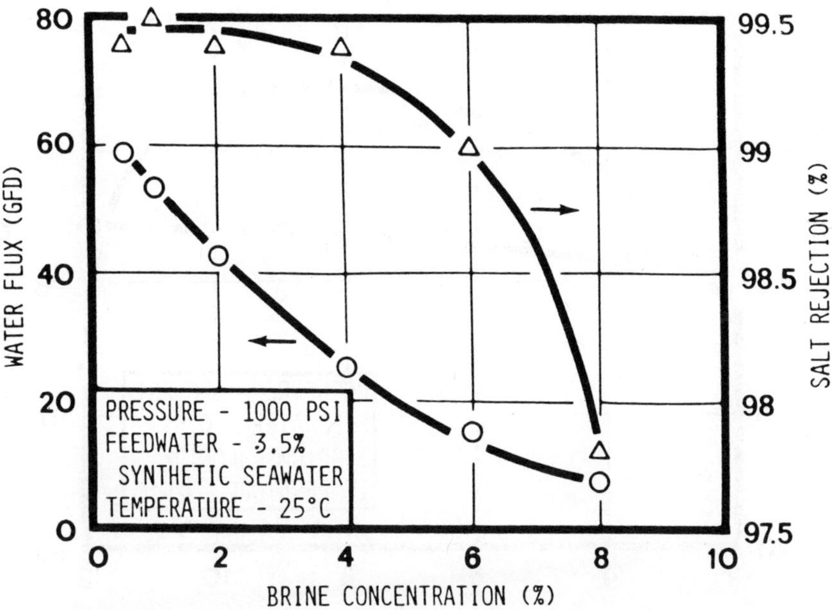

FIGURE 9 Effect of Feedwater Concentration on Flux and Salt Rejection of an FT-30 Thin Film Composite Membrane.

certain point is reached at which salt rejection begins to drop because of the effect of decreased water flux.

Figure 10 examines the effect of feedwater pH on membrane flux and salt rejection. This has to be determined for every new membrane that comes along. Salt rejection will depend on the pH. Water flux may also depend on the pH. In the case of the FT-30 membrane, water flux is basically stable as a function of pH over a range of 5 to pH 12. The highest salt rejection for this membrane occurs at about pH 8.5.

For cellulose acetate and for UOP's PA-300 membrane, maximum salt rejection occurs at about pH 5½. This maximum will vary from one membrane to another. Exactly where this maximum occurs can be advantageous for one membrane over another depending upon the nature of the application. Many groundwaters and well waters, and particularly seawater itself, have naturally occurring pH values in the range of 8 to 8½, in some cases even up to pH 9. With the current trend to polymeric antiscalant addition in place of acidification/sodium hexametaphosphate addition, maximum salt rejection in the pH range of 8 to 8½ may be preferred.

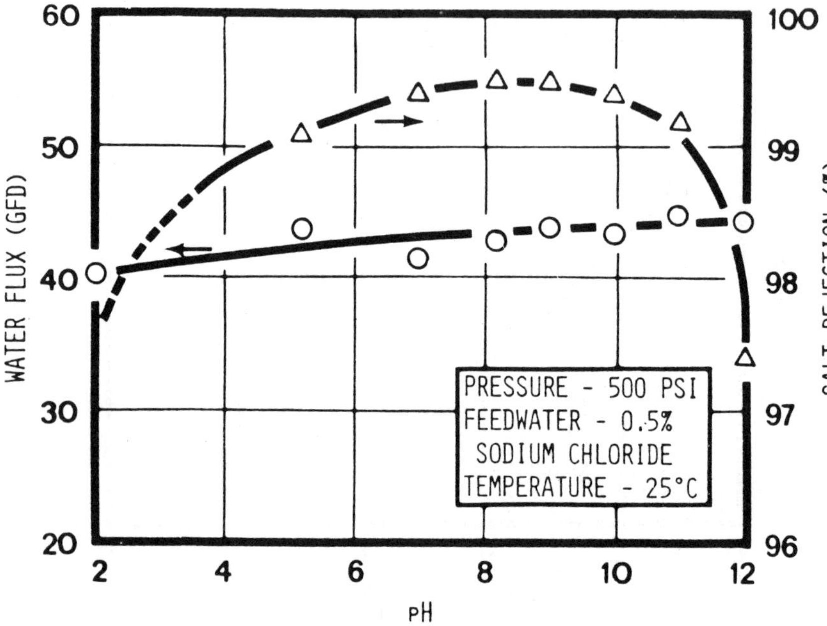

FIGURE 10 Effect of Feedwater pH on Flux and Salt Rejection of an FT-30 Thin Film Composite Membrane.

Figure 11 shows the effect on salt rejection by the feedwater flow rate through a spiral wound membrane module. Because of the effect of concentration polarization, the apparent salt rejection may be quite low at low feedwater channel velocities. However, the membrane salt rejection may actually be high, but the membrane desalinating surface is seeing a much higher salt concentration than exists in the center of the feed channel. All reverse osmosis membrane devices suffer to some extent from concentration polarization, so that the manufacturers of these devices set specifications for minimum feed flow rates.

NEW DIRECTIONS IN REVERSE OSMOSIS MEMBRANES

Seawater desalination is one of the most demanding applications for reverse osmosis, and potentially the most profitable. Continued strong development is expected in this area. This market is principally divided

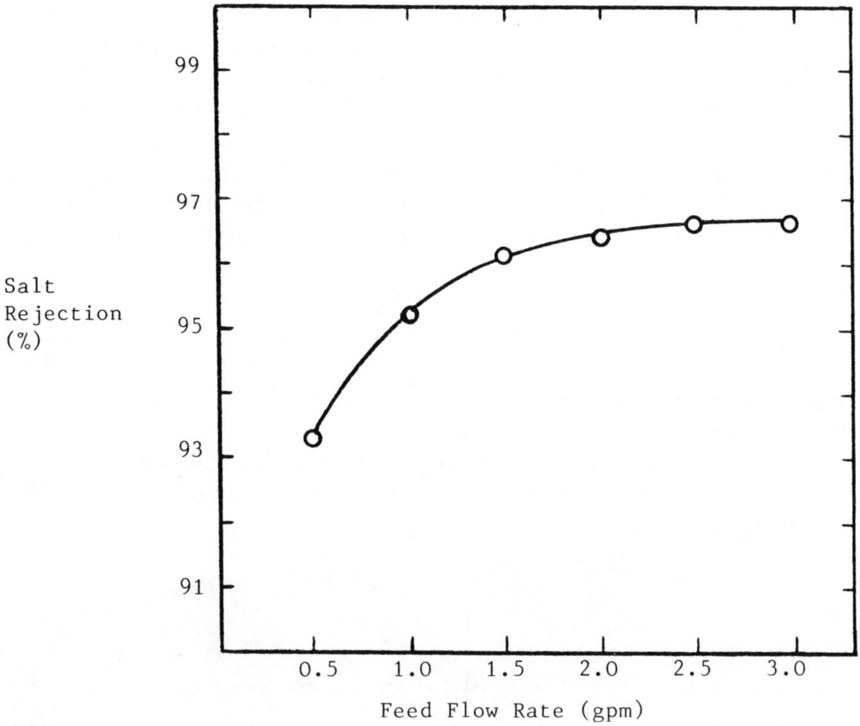

FIGURE 11 Effect of Feedwater Flow Rate on the Salt Rejection of a Membrane Spiral Element.

between the DuPont B-10 Aramid hollow fiber membranes, the Toyobo cellulose triacetate hollow fiber membranes, and the FT-30, PEC-1000, and NS-100 type of thin film composites. Recently, DuPont introduced an improved B-10 hollow fiber capable of operation at 1200 psi and at 40°C. Previous upper limits were 1000 psi and 35°C. Although the flux of this new fiber is initially lower (as much as 20%), the finished module is capable of producing more permeate output and at recoveries as high as 40% on seawater (versus 30% on the previous design) because of the higher operating pressure limit.

UOP is also reported to have announced the availability of an improved spiral element capable of generating 400 ppm product water from Persian Gulf seawater at 30% recovery.

Dow Chemical Company is working on a scaled-up version of their low pressure cellulose triacetate hollow fiber permeators. If they succeed, they will be able to provide a single hollow fiber module capable of producing 250,000 gallons per day. This will represent a quantum leap in the size of reverse osmosis membrane devices.

On the other end of the scale, "loose" reverse osmosis membranes are being developed that allow passage of monovalent salts, but have molecular weight cutoffs of 500 or less. Three such entries are the Nitto Electric NTR 7250 and the FilmTec FT-40 and FT-50 thin film composite membranes. Table III compares performance data for these three membranes.

These loose reverse osmosis membranes are potentially useful for removing salt from a feedstream, rather than concentrating this feedstream. Thus they can be used to "desalt" the concentrate rather than provide a desalted permeate. One potential application, for example, is the desalting of a cheese whey concentrate.

The FT-40 membrane is based on the NS-300 type and is formed from piperazine and trimesoyl chloride to give a crosslinked polyamide of the following structure:

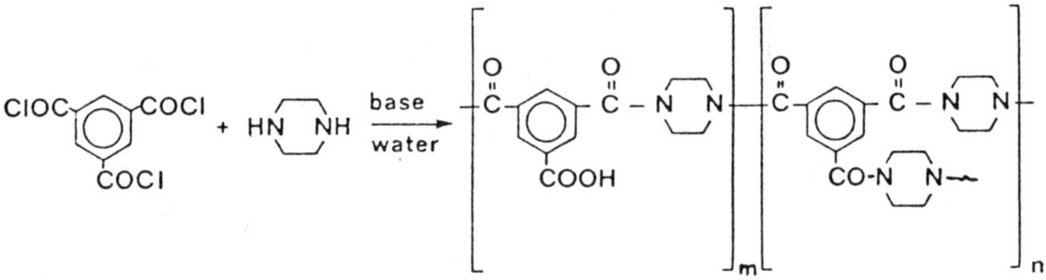

The chemical structure of the NTR-7250 membrane has not been published. Nitto Electric has applied for a patent in which polyvinyl alcohol and piperazine are combined with trimesoyl chloride to give a thin film composite membrane. (29) The salt rejection characteristics of NTR-7250 would fit such a composition, as would the limited pH range of 4–8 due to the probable presence of ester groups (polyvinyl alcohol reaction with trimesoyl chloride).

The highest flux among these three loose membranes is exhibited by the FT-50 membrane. It is capable of being used at pressures of 50 psi and lower. Yet, it still possesses a 500 molecular weight cutoff. This is literally a case of a reverse osmosis membrane operating at ultrafiltration membrane pressures. The FT-50 membrane is still in an experimental prototype stage. Possible applications include final filtration of 18-megohm electronics grade water.

TABLE III

Comparison of Performance Data for "Loose" Reverse Osmosis Membranes

	NTR-7250	FT-40	FT-50 (Exptl)
Flux	35–59 gld/280 psi	18–24 gfd/200 psi	15–20 gfd/50 psi
Salt Rejection			
NaCl	40–60	40–50	30–40
$MgSO_4$	95–99	90–95	90–95
sucrose (M.W. 342)	98–99	98–99	98–99
raffinose (M.W. 504)	99–100	99_+	99_+
pH range	4–8	3–11	2–12
Cl_2 resistance	100 hrs/100 ppm/ soak test	2500 hrs/0.75 ppm/ 200 psi operation	not recommended

MEMBRANE DISTILLATION

A new approach to membrane desalination was recently unveiled by Gore and Associates. This approach uses Gore-Tex microporous polytetrafluoroethylene film for membrane distillation. (30) This film is highly hydrophobic and will not pass liquid water at low pressure. However, it readily passes water vapor. A membrane distillation process can be visualized, Figure 12, whereby Gore-Tex film is positioned between a heated feedwater and a cooler condenser surface. The length of the path of the water vapor from feedstream to condensate stream can be as short as only 0.001 inch (the thickness of a one-mil film of Gore-Tex). The process requires a thermal input into the feedwater to drive the process, but low grade heat sources can be used.

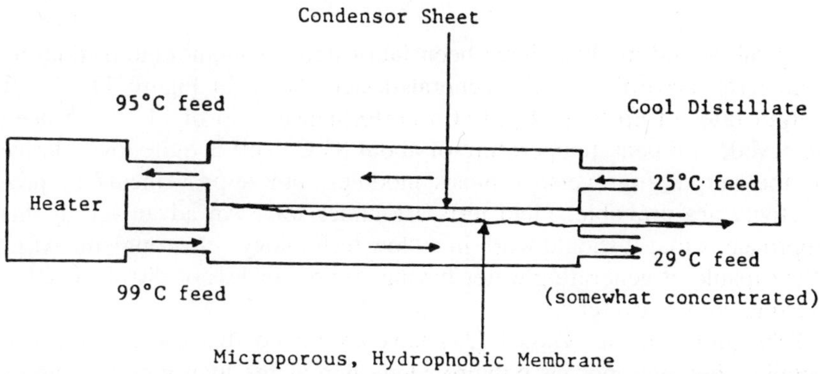

FIGURE 12 Principle of a Continuous Membrane Distillation Device for Seawater Desalination.

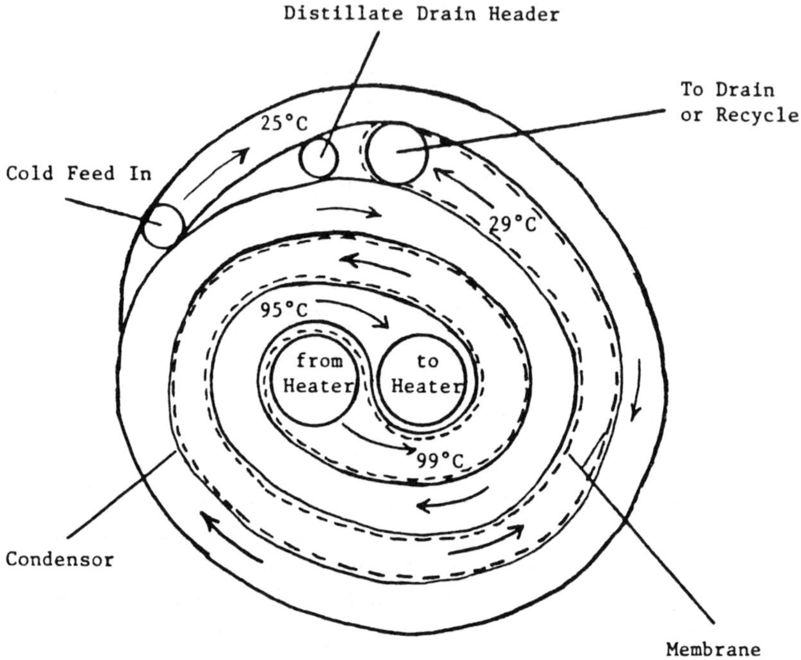

FIGURE 13 Design of a Spiral Wound Membrane Distillation Element.

Spiral wound modules have been fabricated for membrane distillation field tests, according to the general design shown in Figure 13. Water output rates of up to 1.74 gfd at a performance ratio of 14.2 have been achieved, at a peak temperature of about 85°C. This is quite low relative to the output of reverse osmosis modules, but improvements in productivity are probable. Field tests are under way. An advantage of this approach is that it should work in a "low technology" environment. Also, it is capable of generating water having 20 ppm or less of dissolved salts, starting with seawater.

Schneider and van Gassel (31) have examined this process in more detail, using microporous polypropylene film as the hydrophobic, vapor-permeable membrane. At a feedwater temperature of 95°C they achieved productivities as high as 6 to 7 gfd in laboratory tests.

REFERENCES

1. Applegate, L.E., "Membrane Separation Processes," *Chem. Eng.*, *91(12)*, 64–89 (June 11, 1984).
2. "Synthetic Membrane Processes," Ed. by G. Belfort, Academic Press, New York, 1984.
3. Reid, C.E., and Breton, E.J., *J. Appl. Polymer Sci.*, *1*, 133 (1959).
4. Loeb, S., and Sourirajan, S., *Advan. Chem. Ser.*, *38*, 117 (1962).
5. Cannon, C.R., U.S. 3,285,765 (Nov 15, 1966).
6. King, M.W., Hoernschameyer, D.L., and Saltonstall, C.W., Jr., "Cellulose Acetate Blend Membranes," in "Reverse Osmosis Membrane Research," H.K. Lonsdale and H.E. Podall, Eds., Plenum Press, New York, 1972, p. 131.
7. Mungle, C.R., and Fox, R.L., U.S. 4,026,978 (May 31, 1977).
8. Francis, P.S., "Fabrication and Evaluation of New Ultrathin Reverse Osmosis Membranes," available from National Technical Information Service, Springfield, VA 22161, Report No. PB-177083 (1966).
9. Lonsdale, H.K., Riley, R.L., Lyons, C.R., and Carosella, D.P., Jr., "Transport in Composite Reverse Osmosis Membranes," in "Membrane Processes in Industry and Biomedicine," Ed. by M. Bier, Plenum Press, 1971, p. 101.
10. Rozelle, L.T., Cadotte, J.E., Corneliussen, R.D., and Erickson, E.E., "Development of New Reverse Osmosis Membranes for Desalination," available from National Technical Information Service, Springfield, VA 22161, Report No. PB-206329 (1967).
11. a.) Cadotte, J.E., U.S. 4,039,440 (Aug 2, 1977); b.) Rozelle, L.T., Cadotte, J.E., Cobian, K.E., and Kopp, C.V., Jr., "Nonpolysaccharide Membranes for Reverse Osmosis: NS-100 Membranes," in "Reverse Osmosis and Synthetic Membranes," Ed. by S. Sourirajan, available from Publications, National Research Council of Canada, Ottawa, Canada K1AOR6, 1977, p. 249.
12. a.) Cadotte, J.E., U.S. 3,926,798 (Dec 16, 1975); b.) Cadotte, J.E., Kopp, C.V., Jr., Cobian, K.E., Rozelle, L.T., "In Situ-Formed Condensation Polymers for Reverse Osmosis Membranes: Second Phase," National Technical Information Service, loc. cit., Report No. PB-234198 (1974); c.) Cadotte, J.E., Cobian, K.E., Forester, R.H., Petersen, R.J., "Continued Evaluation of In Situ-Formed Condensation Polymers for Reverse Osmosis Membranes," ibid., Report No. PB-253193 (1976).
13. a.) Cadotte, J.E., U.S. 4,259,183 (Mar 31, 1981); b.) Cadotte, J.E., King, R.S., Majerle, R.J., and Petersen, R.J., *J. Macromol. Sci-Chem.*, *A15(5)*, 727 (1981).
14. a.) Cadotte, J.E., U.S. 4,277,344 (Jul 7, 1981); b.) Cadotte, J.E., Petersen, R.J., Larson, R.E., and Erickson, E.E., *Desalination, 32*, 25 (1981); c.) Cadotte, J.E., and Petersen, R.J., "Thin-Film Composite Reverse Osmosis Membranes: Origin, Development, and Recent Advances," in ACS Symposium Series No. 153, "Synthetic Membranes: Volume I, Desalination" Ed. by A. Turbak, American Chemical Society, Washington, D.C., 1981, p. 305.
15. "Why Hollow Fiber Reverse Osmosis Won the Top CE Prize for DuPont," *Chem. Eng., 78(24)*, 53 (Nov 29, 1971).
16. Richter, J.W., and Hoehn, H.H., U.S. 3,567,632 (Mar 2, 1971).
17. Dance, E.L., Davis, T.E., Mahon, H.I., McLain, E.A., Skiens, W.E., and Spano, J.O., "Development of Cellulose Triacetate Hollow Fiber Reverse Osmosis Modules for Brackish Water Desalination," Available from National Technical Information Service, loc. cit., Report No. PB-208786 (Dec 1971).
18. Davis, T.E., "One Pass Seawater Desalination Hollow Fine Fibers," ibid., Report No. PB-234101 (Jun 1974).
19. Kunisada, Y., Kaneda, H., Hirai, M., and Murayama, Y., *Desalination, 30,* 337 (1979).
20. Spatz, D.D., "Reverse Osmosis: the Mechanism and Application to Dialysis," *Medical Instrumentation, 8(3),* 209 (1974).
21. Vos, K.D., Burris, F.O., Jr., and Riley, R.L., *J. Appl. Polymer Sci., 10,* 825 (1966).
22. Spatz, D.D., and Friedlander, R.H., "Rating the Chemical Stability of U.C. RO/UF Membrane Materials," *Water & Sewage Works, Feb 1978*, p. 36.

23. Sudak, R.G., Bott, J.B., Fox, R.L., Gagner, S.J., and McKee, M.E., "Development of Reverse Osmosis Membrane for Seawater Desalination," Available from National Technical Information Service, loc. cit., Report No. PB80-182090 (September 1979).
24. a.) Wrasidlo, W.J., U.S. 4,005,012 (Jan 25, 1977); b.) Riley, R.L., Fox, R.L., Lyons, C.R., Milstead, C.E., Seroy, M.W., and Tagami, M., *Desalination, 19,* 113 (1976).
25. Hickman, C.E., Jamjoom, I., Riedinger, A.B., and Seaton, R.E., *Desalination, 30*, 259 (1979).
26. Kurihara, M., Watanabe, T., and Inoue, T., U.S. 4,366,062 (Dec 28, 1982).
27. Chen, J.-Y., Kurihara, M., and Pusch, W., *Desalination, 46,* 379 (1983).
28. Bhattacharyya, D., Jevtitch, M., Ghosal, J.K., and Kozminski, J., "Reverse Osmosis Membrane for Treating Coal-Liquefaction Wastewater," *Environ. Prog., 3(2),* 95 (May 1984).
29. Kamiyama, Y., Yoshioka, N., and Nakagome, K., Ger. Offen. DE 3,129,702 (Feb 11, 1982).
30. Core, D.W., "Gore-Tex Membrane Distillation," Tech. Proc. 10th Ann. Conf. Water Supply Improvement Association, Vol III, Honolulu, Hawaii, July 25–29, 1982.
31. Schneider, K., and van Gassel, T.J., "An Energy Efficient Membrane Distillation Process," paper presented at Europe-Japan Congress on Membranes and Membrane Porcesses, Stresa, Italy, June 18–22, 1984.

Current Aspects of Membrane-based Separation of Gases

W. J. KOROS AND D. R. PAUL

Center for Energy Studies and Department of Chemical Engineering, The University of Texas at Austin, Austin, Texas 78712

Important existing types of commercial applications of membranes for the separation of gas mixtures are reviewed. The strengths and weaknesses of current membrane processes are discussed briefly relative to more traditional competitive technologies for such applications. The major technical developments leading to the current popularity of membranes are traced historically. Fundamental principles governing the operation of gas separation membranes are presented. Finally, optimum directions for research to develop advanced membranes for today's and future markets are then suggested, based on these principles.

INTRODUCTION

The energy efficiency and simplicity of membrane separation devices make them extremely attractive for solution of fluid phase separation problems. The ability of ideal membrane processes to selectively pass one component in a mixture, while rejecting others in a continuous steady-state manner, defines the perfect separation device. A substantial literature related to gas separation indicates, however, that numerous factors must be considered in achieving a commercially successful membrane-based process. Nevertheless, recently membranes have been shown to act as remarkably effective separating devices (1–11). By clever engineering, their productivities have been increased dramatically, and they are strongly competitive with, if not clearly superior to, more traditional chemical engineering separation approaches for several applications involving gas treatment.

Due to space limitations, the present discussion focuses only on polymeric membrane processes for gas separations. A related approach,

Presented at the Symposium on Synthetic Membranes at the Sixteenth Annual International Symposium sponsored by the Michigan Molecular Institute, Midland, Michigan, August 19–24, 1984.

termed "supported liquid membranes," involves the use of low-molecular-weight fluids in the pores of a support membrane. Liquid membranes offer the potential for high selectivity under certain conditions. A still-unresolved drawback of liquid membrane systems, however, lies in the tendency of the liquid components of the membrane to migrate from the membrane support into the external surrounding phases. These systems have been reviewed by Scholtz, Goddard, and Suchdeo (12) and are not treated further in the following discussion.

APPLICATION EXAMPLES

Applications of membrane-based gas separation technology tend to fall into three major categories at the present time:

1. H_2 separation from a wide variety of slower permeating supercritical components such as CO, CH_4, and N_2 (1–3, 7, 10, 11, 13)
2. Carbon dioxide and water separation from natural gas (1–6, 9, 11, 13, 14)
3. O_2 or N_2 enrichment of air (14–16)

The order of the various types of applications given above provides a qualitative ranking of the relative ease of performing the three types of separations. The extraordinarily small molecular size of H_2 makes it extremely permeable and easily collected as a permeant compared to the other, more bulky supercritical gases indicated (17). Perhaps surprisingly, it is difficult to separate H_2 from CO_2, although this latter gas is clearly much larger in size. Because of its low condensibility, H_2 has a very low solubility in membranes (18, 19). Therefore a more soluble, lower diffusivity gas such as CO_2 may have a steady-state permeability comparable to that of H_2, since the product of solubility and diffusivity determines the permeability of a component through the membrane.

A study by Monsanto involving an H_2 recovery permeation unit on an operating 100-million-gal/yr low-pressure methanol plant illustrates the value of the gas separator as a cost-saving unit operation (7). The study also emphasizes the fact that H_2 and CO_2 are not easily separated. In this particular case, however, the relative inseparability of the pair turns out to be an *advantage* since both components can be split from a purge stream and recycled to the synthesis gas reactor.

The permeant from each permeator is recycled to the synthesis reactor, and the nonpermeant from the last separator is sent to the fuel header to be burned. For a feed side pressure of 682 psig and a bore side pressure of 380 psig, the purge gas is reconstituted to almost 80% H_2 and depleted somewhat in CO_2 because the permeabilities of H_2 and CO_2 are not identical. The separators are reported to recover approximately half of the available H_2 and CO_2, which had previously been sent to the fuel header.

Reportedly a net 2.4% increased production rate of methanol can be achieved by elimination of this raw material waste. The slightly slower permeation rate of CO_2 relative to H_2 requires the addition of a small amount of CO_2 to the recycle stream to maintain the 3-to-1 stoichiometry of H_2 to CO_2 required by the overall synthesis reaction (7). Accounting for this supplemental CO_2 and a small debit for the fact that the entire purge stream can no longer provide fuel value, an estimate of the net advantage of the permeator installation was reached by the Monsanto engineers. Even with the above debits, the permeator installation permitted a 13% reduction in incremental costs to produce each additional gallon of methanol from the purge stream components that had been burned previously (7).

ENHANCED OIL RECOVERY, CO_2 RECLAMATION

Enhanced oil recovery (EOR) operations often involve injection of high-pressure (2,000 psig) CO_2 into a reservoir. Subsequent removal of CO_2 occurs along with oil and light gases (primarily methane) from wells drilled at a distance from the primary injection point (see Fig. 1).

The reasonably high solubility of CO_2, H_2S, and water in membranes at low partial pressures, coupled with the relatively low solubility and diffusivity of the bulky supercritical methane molecule, has made possible this second class of separations. The primary difficulties one encounters in this class of systems arise from possible loss in permselectivity of the dense selective layer and compaction of the open cell support structure at high pressures as a result of plasticization by the permeating components. The gaseous products can be separated into a CO_2-rich stream (95% CO_2) and used for reinjection, and a CH_4-rich stream (98.5% CH_4) and used for pipeline gas. Alternatively, the stream can be conditioned to only 90% CH_4 and used as a field fuel.

A study by researchers at Shell (6) considers the merits of membrane technology for these CO_2 recovery applications relative to other competitive technologies such as physical and chemical solvent scrubbing and cryogenic treatment. The Shell study was based on the permeability and selectivity properties of commercial asymmetric cellulose acetate membranes. The permeability and selectivity properties of the membranes

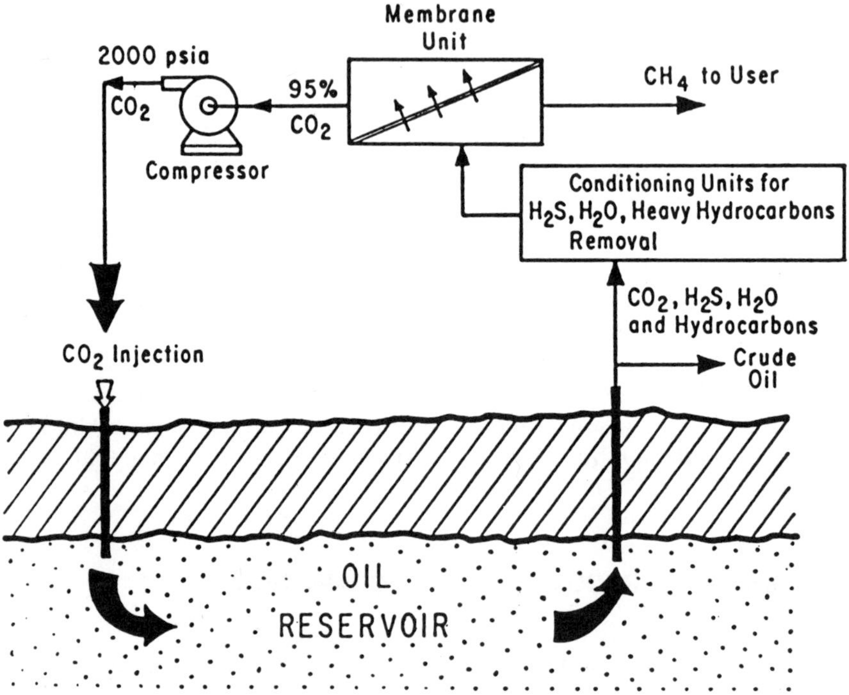

FIGURE 1 Schematic representation of enhanced oil recovery application of membranes for CO_2/CH_4 separation from gases recovered from a production well after injection of CO_2.

were assumed to be independent of gas feed pressure and composition as a first approximation. There is considerable evidence that at high CO_2 partial pressures, this assumption may break down for pure CO_2 feeds because of plasticization effects.

It has been reported, however, that the plasticization effects appear to be less severe in actual mixed gas situations in the presence of substantial methane partial pressures (6). Due to the ambiguity of published data on this topic, the neglect of the plasticization effect by the Shell group seems reasonable at the present time. The behavior of polysulfone and cellulose acetate (in the absence of the ill-defined plasticization issue) is similar to a first approximation.

The results of the study are, therefore, reasonably representative of both types of existing membrane systems. The study permits insights into the strengths and weaknesses of the current generation of membranes compared to chemical and physical solvent treatment systems for these gas purification applications. The Shell group suggests operating regimes in which membranes are clearly superior to other technologies. It also identifies regimes where membranes have problems competing with traditional technology because of insufficient selectivity and permeability of current membrane materials. This lack clearly emphasizes the need for the continued development of membrane materials, since membrane-based processes should, in principle, be attractive over a very large range of operating conditions. This topic is discussed in a later section.

In the Shell study credit was taken for heavy hydrocarbons that are retained in the residual gas by membrane processes and that are typically lost along with the CO_2 in physical solvent systems. This benefit can be substantial in some cases for these valuable components. The Shell researchers concluded:

> The membrane processes are competitive for separation of gases containing high CO_2 concentrations into residual gases containing substantial CO_2, say 10% volume. Membrane processes appear to have a greater sensitivity to the CO_2 partial pressure ratio (inlet CO_2 partial pressure/nonpermeant CO_2 partial pressure) compared to physical solvent processes. However, further optimization of the membrane processes would reduce this difference. Membrane processes are not competitive with the chemical solvent processes for gases with low CO_2 concentrations. This shows that the membrane applications for CO_2 recovery from flue gases are not viable with today's technology.

Higher permeability, more selective membranes could alter the last conclusion as the active research programs in this area yield second-generation materials superior to cellulose acetate and polysulfone. Applications in which membranes are used as "topping units" for high-CO_2-content streams are extremely attractive even with current membrane materials. Another useful discussion of EOR applications of membranes has been presented by Schendel, Mariz, and Mak (11). These authors make comparisons with the Ryan-Holmes low-temperature distillation process. They found slight cost advantages associated with the membrane approach. More significant, however, was the important operating flexibility of membrane systems. They note:

> The real advantage with membrane use is flexibility and modular design. A reliable forecast of ultimate gas production is not required, and will not significantly affect the economics of the plant. With a pure distillation scheme, a plant must be built to accommodate forecasted gas production. If the plant is oversized, this will result in extra capital being spent for unused capacity; if the plant is undersized, inability to process gas may limit oil production.
>
> Because of extremes in gas volume processed (turn-down ratio), the

design basis in this case represents only half the anticipated peak volume. A second distillation plant will be required. With membranes, however, additional membrane capacity at the front end can be added as needed without additional equipment required in the distillation section. Analyses of several associated gas forecasts indicate the exit gas from the first stage of membranes to be remarkably stable in composition and flow rate. This flexibility is not shown in a single case economic comparison.

O_2/N_2 SEPARATIONS

The most difficult of the three types of separations mentioned above is the last one involving O_2 and N_2. The market for O_2-enriched air for medical and furnace applications is considerable. Moreover, N_2-enriched air for blanketing of fuels and stored foods to provide nontoxic, non-residual protection from fire and pests that require O_2 to exist is an interesting possibility.

Currently available polymer membranes to separate O_2 and N_2 have only moderate selectivities (20). One can understand this situation in terms of the combined diffusion and solubility contributions to component permeabilities. The size and shape (and hence diffusivity) of O_2 and N_2 are quite similar; moreover, the solubility of the pair in most membranes is similar. The ratio of the permeabilities of the two components tends to be similar, and, hence, the ideal separation factor is low.

Nevertheless, because of the importance of the problem, processes have been designed that are able to produce economical supplies of O_2- and N_2-enriched air for commercial applications. Monsanto has recently announced plans to use its Prism® modules, based on polysulfone hollow fibers, to provide N_2 as a blanketing agent for fuels in tanker shipments (16). Japan has a very active program underway to develop high-flux separation membranes for a wide spectrum of O_2-enrichment applications, including the improvement of burner efficiencies. A program at General Electric in the 1960s demonstrated the technical feasibility of such processes using flat, low-surface-area plate-and-frame modules composed of ultrathin membranes. This project, however, has been terminated, and the technology is currently being used only to supply O_2-enriched air for personal medical applications (14).

The most probable reason for the slow development of the O_2/N_2 membrane systems is the strong competition encountered from existing highly optimized cryogenic processes. For large-scale operations, cryogenic processes become progressively more economical, while each additional unit of capacity requires an exactly corresponding additional unit of membrane area to perform the separation. Thus, membranes are most competitive in small and intermediate-scale O_2/N_2 operations in which flexibility of operation is valuable.

MODULE AND MEMBRANE IMPROVEMENTS

General

Two serious problems faced by pioneers in gas separation were the low selectivities and the rather low permeation fluxes observed for most membranes. The low-flux problem arose because membranes had to be thick (at least 1-mil) to avoid pinholes, which destroy selectivity due to almost indiscriminate passage of both feed components by Knudsen or viscous flow.

Discovery of the extraordinarily high permeability of silicone polymers (see Table I for comparisons to other typical rubbery and glassy polymers)

TABLE I

Comparison of permeabilities of important product gases in typical rubbery and glassy polymers at 25°C–30°C and 1 atm [23].

Polymer	Repeat Structure	P_{H_2}*	P_{CO_2}*	P_{O_2}*
	Rubbery Polymers			
Silicone Rubber		550	2700	500
Natural Rubber		49	131	24
Polychloroprene (Neoprene®)		20	22	4
	Glassy Polymers			
Poly(Ethylene Terephthalate) (Mylar®)		0.6	0.10	0.03
Aromatic Polyether Diimide) (Kapton*110)		1.5	0.27	0.15
Polycarbonate (Lexan®)		12	5.6	1.4

* Permeability units are Barrers. 1 Barrer = 10^{-10} [cc(STP) cm]/[cm^2 sec cm Hg]

spurred a renewed interest in gas separation aimed at O_2/N_2 separation from air (21–23). Unfortunately, the O_2/N_2 selectivity of silicones is rather low. Moreover, even with silicone polymers of 1-mil thickness, very large membrane areas were required. Stern showed that membrane separations were not competitive with cryogenic processes using 1-mil silicone rubber membranes even if the feed air was compressed considerably to reduce installed membrane cost (24). The development of casting techniques for ultrathin membranes composed of (150 Å) silicone rubber/polycarbonate copolymers permitted formation of pinhole-free selective membranes with thicknesses around 1,000 Å by laminating multiple layers (15). Such membranes offered flux increases of as much as 50- to 100-fold compared to their 1-mil silicone rubber counterparts. The membranes were typically supported in a plate-and-frame fashion on a porous substrate, so that concentration of a large membrane area in a small enclosure was not feasible.

The so-called hollow-fiber and spiral-wound approaches described below have been adopted in the gas separation field (as in the reverse osmosis and ultrafiltration fields) as the most viable solution to the problem of incorporating large amounts of surface area into compact module volumes to obtain high area densities. Permeation area densities can, for example, be increased roughly 30-fold simply by replacing the plate-and-frame flat membrane approach with 200-μ outside diameter (OD) hollow fibers packed at a 50% void factor in a cylindrical shell, such as shown in Fig. 2.

As shown in Fig. 2, the structure of the hollow fibers is complex, containing essentially pore-free skins of 1,000 Å to 2,000 Å thickness

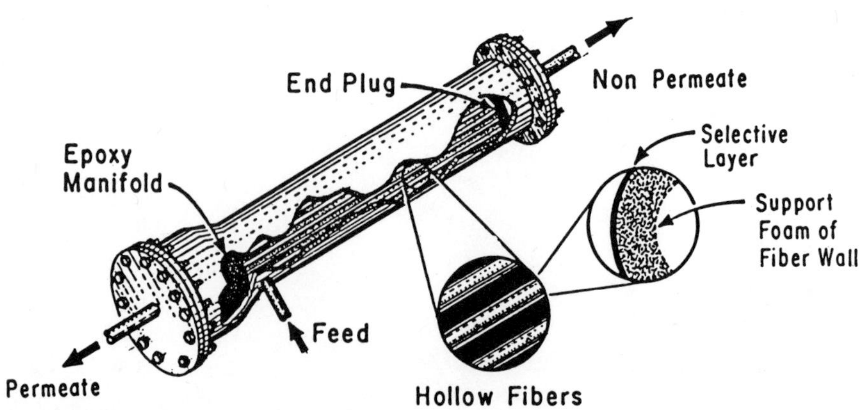

FIGURE 2 Cutaway view of a typical hollow-fiber gas separation module.

supported in a highly porous open cell foam support. The reduction in effective membrane thickness from 1 mil to 1,000 Å increases flux more than 200-fold under equivalent operating conditions.

In the Monsanto module design, the hollow-fiber bundle is closed at one end with an epoxy plug or similar device (as shown in Fig. 2). Feed enters at the side of one end of the module and flows axially along the fibers, ultimately exiting as nonpermeant at the end opposite where it entered. Permeant diffusion through the hollow-fiber wall into the bore moves countercurrently to the stream in the shell side and is ultimately collected in the chamber where all of the open ends of the fibers terminate.

Alternatively, in the Dow design, feed and nonpermeant enter and exit at opposite ends of the shell, while permeant product gas is removed from both ends of the module. This latter design permits the use of longer or smaller diameter fibers without serious pressure buildup in the bore (25).

The early flat membrane configuration typically provides area densities of only about 300 ft^2/ft^3 of module volume (15), while the 200-μ diameter fibers give densities of more than 9,000 ft^2/ft^3 of module volume. Clearly, even higher area densities can be achieved using smaller diameter fibers and higher packing densities. Studies have been reported using 36-μ diameter fibers in test cases with low-flux polyester fibers (26, 27). For the high-flux fiber currently in use, however, dimensions in the range of 100-μ to 500-μ OD appear to be favored on the basis of the information available in the open literature (28–31).

A second type of high-surface-area module design also relys on asymmetric membranes, but not in fiber form. This module type is typically referred to as "spiral wound" and is illustrated in Fig. 3. A sandwich consisting of a porous backing material (e.g., Dacron® felt) placed between the two halves of a folded continuous asymmetric membrane is assembled with a separator grid (e.g., polypropylene screen) on the top membrane half. The porous backing material is wrapped around the perforated mandrel to provide an easy pathway for gas to the mandrel holes. The sandwich is then sealed along its edges with a suitable epoxy at three sides. The fourth side terminates at the central perforated mandrel. Finally, the assembly is simply spirally wound and loaded into a suitable cylindrical chamber such as that shown in Fig. 3.

A feed stream enters at one end and moves through the passageways provided by the polypropylene grid separator. The nonpermeant then exits at the end opposite where it entered. During passage through the module, permeant diffuses across one of the two membranes in the spirally wound sandwich and flows in the porous backing material inwardly toward the open end of the sandwich terminating at the perforated mandrel. The permeant then accumulates in the central collection tube

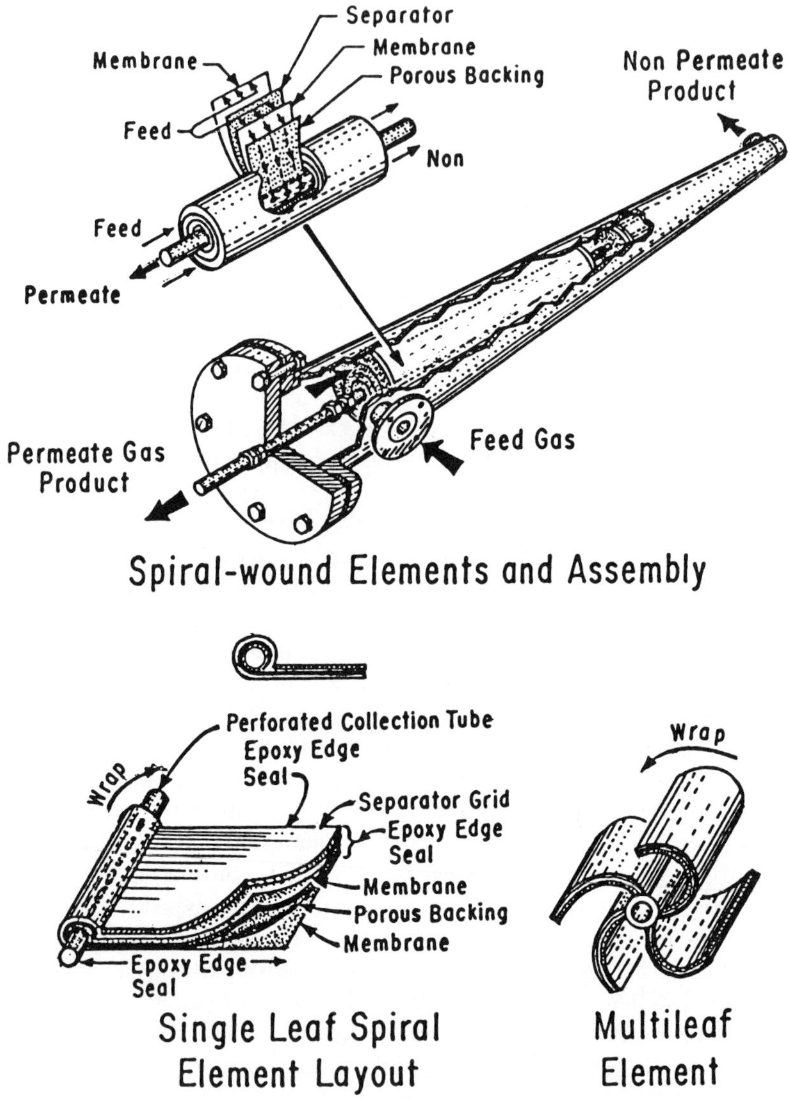

FIGURE 3 Spiral-wound elements and assembly. The use of the multileaf element avoids long pathways for permeate through the porous backing material to minimize pressure buildup in the permeate channel.

and is delivered as product. As a means of incorporating more area in a compact volume, multileaved cartridges can be used without requiring excessively long flow paths of permeant to the collection tube (32).

Coupling the higher permeation area density resulting from the spiral-wound or hollow-fiber approach with the much smaller membrane resistances resulting from the asymmetric membrane approach produces flux densities that are on the order of ten thousand times higher than those available in the early days of gas separation in the 1960s. Consideration of these impressive strides makes it clear why this technology has gained so much attention from separation engineers recently.

The degree of perfection of the membrane skins that perform the actual separation must be very high. Pores of even 10 Å to 50 Å present at an area fraction of even 10 ppm can destroy the permselectivity of a membrane that must discriminate between gas molecules with characteristic dimensions of 3 Å to 5 Å. Monsanto introduced a process to posttreat asymmetric fibers with a minuscule amount of highly permeable "stop leak" material such as silicone rubber. This treatment eliminates most of the Knudsen and viscous contributions to flow through the membrane and produces a high-flux membrane with selectivity approaching that of true dense films (33).

Cellulose acetate (CA) and modified cellulose acetates of various degrees of acetylation comprise a second major commercial material used by a number of companies in the field. Dow, Separex, Environgenics, and Grace are involved in either hollow-fiber (Dow) or spiral-wound module approaches (Separex, Environgenics, and Grace) to membrane packaging using these polymers (4–6, 34). Little detailed information is available concerning the performance of cellulose acetate in actual gas separation applications, but it appears on the basis of the available literature to be roughly equivalent to polysulfone (9–11). Recent evidence suggests, however, that the CA may be somewhat less tolerant to high temperatures and elevated CO_2 partial pressures than the corresponding polysulfone membranes (35); however, no truly "head-to-head" comparisons of the two membrane types have been reported.

An interesting alternative approach to membrane formation has been the basis for reverse osmosis membranes marketed by Film Tech, UOP, and Albany International. These composite membranes are essentially the mirror images of the Monsanto composites. In these alternative cases the membrane support is highly porous, and a thin selective coating (~1,000 Å to 2,500 Å) is formed on the porous support by interfacial polymerization (14, 23, 36–38).

This approach is the reverse of the Monsanto "stop leak" technique since the coating in the interfacial polymerized case forms the actual separating layer and the highly porous sublayer acts simply as a support. The interfacial polymerization approach is clearly attractive, since a selective coating can be formed from a wide range of monomers displaying a tremendous spectrum of physical and chemical properties. The durability of such composite membranes is reported to be quite good in reverse osmosis applications where they were first developed (36). Clearly, applications in gas separation service involve rather moderate sorption levels, so that swelling stresses likely to dislodge such coatings should be tolerable. Therefore, this approach is likely to see more emphasis in the future, but it is not currently being used in any commercial membranes for gas separation.

Fundamentals

Treatment of simple one-dimensional diffusion through a flat film is sufficient to describe the fundamental principles controlling membrane-based gas separation processes. The effects of membrane asymmetry will be neglected since the process of permselection occurs in a thin dense layer of effective thickness, ℓ, at the membrane surface. In such a case, the expression for the local flux of a penetrant at any point in the dense layer can be written as shown in Eq. (1) (39):

$$N = -D\frac{dC}{dx} \tag{1}$$

where the diffusion coefficient, D, may be a function of local concentration, C, in the membrane. The diffusion coefficient appearing in Eq. (1) may be expressed in terms of a preexponential factor, D°, and an exponential term that depends on the activation energy for the diffusion process, E_D (39).

$$D = D^\circ \exp[-E_D/RT) \tag{2}$$

The activation energy is the average energy that must be localized next to the penetrant to generate a sufficiently large opening to permit the molecule to execute a jump. If concentration dependence of the diffusion coefficient exists, both D° and E_D may be functions of the amount of penetrant present. At low concentrations the activation energy should be related strongly to the minimum cross-section of the penetrant, since this defines the critical opening size that must exist for a jump to occur. The large effect on molecular mobility caused by differences in penetrant size and shape is illustrated in Fig. 4 for a typical rubbery and glassy polymer, respectively (19, 40).

The larger nonspherical penetrants tend to approach a plateau diffusivity because the cross-sectional area of the molecule as well as its volume determines its ability to find a gap of sufficient size to permit a jump. Presumably, nonspherical penetrants move in a somewhat oriented fashion, and the individual jump lengths may be only fractions of the total

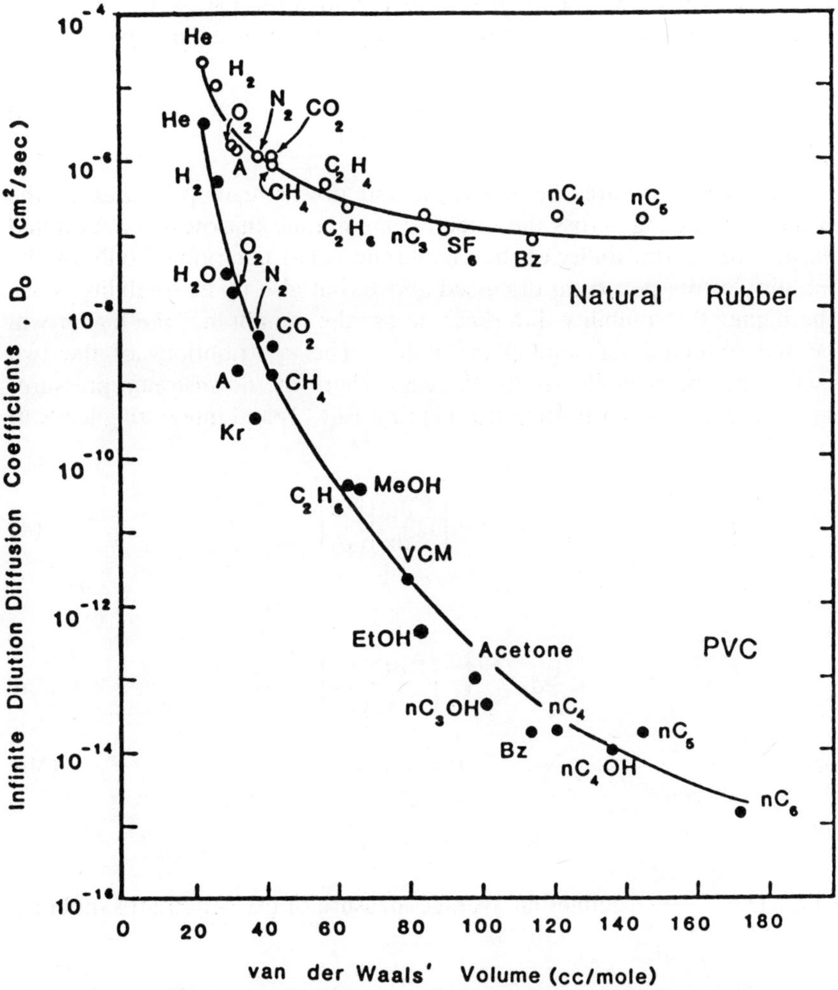

FIGURE 4 Diffusion coefficients for a variety of penetrants in natural rubber and rigid poly(vinyl chloride) at 25°C–30°C.

length of the molecule. Associated with the more highly restrictive glassy environment comes a greater ability to perform size and shape discrimination among penetrants. This difference between the glassy and the rubbery state is illustrated clearly by the much larger spread in diffusion coefficients in the glassy polymer compared to the rubbery one. Recognition of the preceding facts often leads to the generalization that low membrane permeability is necessary for high selectivity. While it is true that the two generally correlate, there appear to be attractive exceptions to the rule that are worthy of serious investigation (41, 42).

The permeability of a membrane to a penetrant is defined by Eq. (3):

$$P = \frac{N\ell}{[p_2 - p_1]} \tag{3}$$

where p_2 and p_1 are the upstream and downstream pressures of the component acting across the effective membrane thickness, ℓ. As noted earlier, the permeability of the membrane is not determined only by the mobility of the penetrant discussed above, but also by its solubility, since the higher the solubility difference across the membrane, the higher will be the observed flux and permeability. The contributions of the two factors can be seen clearly for the case where the downstream pressure, p_1, is negligible. Substituting Eq. (1) into Eq. (3) and integrating leads to Eq. (5):

$$P = -\left[D\frac{dC}{dx}\right]\left[\frac{\ell}{p_2}\right] \tag{4}$$

so that

$$\int_0^{\ell} \frac{P}{\ell} dx = \int_0^{C_2} \left[\frac{DdC}{C_2}\right]\left[\frac{C_2}{p_2}\right] = \overline{ds} \tag{5a}$$

or

$$P = \overline{ds} \tag{5b}$$

where $\bar{D} = \dfrac{\int_0^{C_2} DdC}{\int_0^{C_2} dC}$ defines an average measure of the penetrant's mobility in the membrane between the upstream concentration, C_2, and downstream concentration, $C_1 \simeq 0$ (43). The parameter, $\bar{S} = C_2/p_2$, equal to the secant slope of the sorption isotherm evaluated at the upstream conditions, is a measure of the solubility of the penetrant in the membrane. The solubility factor in Eq. (5) is thermodynamic in nature and is determined by polymer-penetrant interactions and the amount of excess interchain gaps existing in the glassy polymer (43). Since the steady-state permeability, P, and the apparent solubility, $\bar{S}$, can be measured independently, $\bar{D}$ can be calculated by rearranging Eq. (5b). The average diffusion coefficient, $\bar{D}$, is kinetic in nature and is largely determined by polymer-penetrant dynamics as discussed in the context of Eq. (2).

When the downstream pressure, p_1, is negligible compared to the upstream pressure, p_2, the separation factor, α_{AB}, defined by Eq. (6) can be related to the ratio of the permeabilities of components A and B as shown in Eq. (7) (44):

$$\alpha_{AB} = [y_A/y_B][x_A/x_B] \tag{6}$$

$$\alpha^*_{AB} = P_A/P_B \tag{7}$$

where x_i's and y_i's are the local mole fractions of components A and B at the upstream and downstream membrane faces, respectively. The superscript * in Eq. (7) indicates that α^* is the so-called ideal separation factor that applies under the conditions described above. The ratio of the component permeabilities in such a case provides a useful measure of the intrinsic permselectivity of a membrane material for mixtures of A and B.

According to Eq. (7), the ratio of permeabilities is given by Eq. (8) after substitution from Eq. (5) for components A and B:

$$\alpha^*_{AB} = P_A/P_B = [\bar{S}_A/\bar{S}_B][\bar{D}_A/\bar{D}_B] \tag{8}$$

Solubility Selectivity ↙ $[\bar{S}_A/\bar{S}_B]$; Mobility Selectivity ↘ $[\bar{D}_A/\bar{D}_B]$

Eq. (8) demonstrates that the ideal separation factor can be partitioned into a so-called solubility selectivity, $[\bar{S}_A/\bar{S}_B]$, and a mobility selectivity, $[\bar{D}_A/\bar{D}_B]$.

In the absence of strong interactive effects between permeating components, the pure-component permeabilities of the two penetrants in the mixture can be used in Eq. (7). For rubbery polymers at low penetrant partial pressures, the assumption of noninteraction appears to be satisfactory (45–46). The assumption of strict noninteraction between permeating components does not, however, appear to hold in general for glassy polymer membranes (4, 47–51).It also has been shown that plasticization of both rubbery (52) and glassy (35) polymers can occur at high penetrant activities.

On the basis of a recent study of the permeabilities of Kapton® polyimide to CO_2/CH_4 mixtures, we expect that the permeability ratio of CO_2 over CH_4 in a mixture can be approximated to within about 20% by using the respective pure-components permeabilities for systems in which the permeability of the pure components is essentially constant or decreases slightly over the partial pressure range of interest (48). Consequently, for the present general discussion, pure-component values are used in Eq. (8) in the following sections for discussing this important gas system since the multicomponent data available currently are too limited to permit focusing discussion on them at the present. Representative CO_2 permeabilities for several glassy polymers at 35°C are plotted as a function of the upstream penetrant pressure in Fig. 5 (53). Except for cellulose acetate, the permeabilities decrease monotonically with upstream pressure. The behavior of cellulose acetate is addressed in a later section in terms of plasticization arguments.

Examination of Table II and Fig. 5 supports the common observation that low permeability and high selectivity are generally related. For example, the CO_2 permeability of poly(phenylene oxide) (PPO) is more than twelve times higher than that of polysulfone, but its selectivity is roughly one-half that of polysulfone. Kapton®, on the other hand, is more than twenty-five times less permeable to CO_2 than polysulfone is, but its selectivity is more than twice as high as that of polysulfone.

Data such as those shown in Table II and Fig. 5 are typical of trends one finds if a random Edisonian screening approach is taken to select polymers for gas separation membranes. On the other hand, such results are *not* universally observed as shown by the results in Table III, where very impressive H_2 and CO_2 permeabilities and H_2/CH_4 and CO_2/CH_4 ideal separation factors are reported.

The data in this table reflect the potential value of rationally designing polymer structures to achieve desirable gas separation characteristics rather than relying on existing mass-produced engineering resins. There are, of course, problems with relying on specialty polymers since cost factors are considerable and developing expertise in handling such materials may take considerable time and effort not required when using standard materials such as polysulfone and cellulose acetate. The standard polysulfone and cellulose acetate materials whose properties are shown in Table II have served as very effective first-generation membrane materials; and the novel polymers A, B, C, and D, which are discussed in more detail below, can be viewed as logical extensions of such first-generation membrane materials.

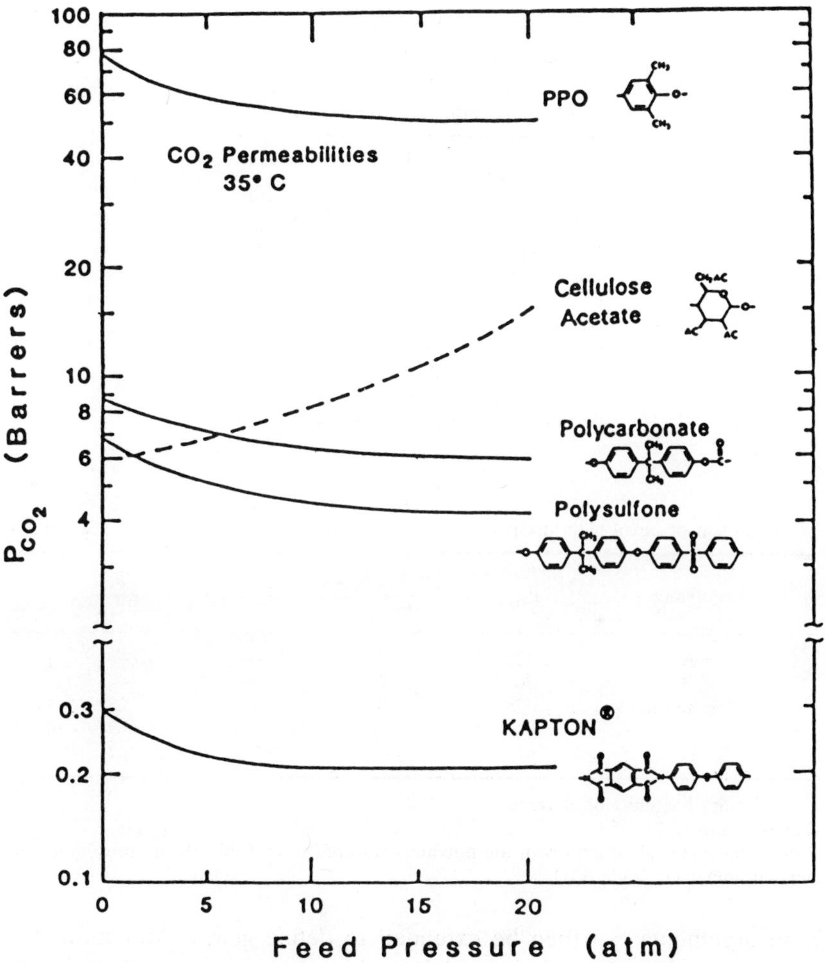

FIGURE 5 Pressure dependence of CO_2 permeability in a variety of glassy polymers at 35°C. The cellulose acetate data are estimated from a number of sources [53].

SOLUBILITY AND DIFFUSIVITY CONSIDERATIONS

It is useful to consider the separate penetrant solubility and mobility contributions, $\bar{D}$ and $\bar{S}$, for the various polymers shown earlier in Fig. 5 to improve one's understanding of why novel materials such as those indicated in Table III may show both high selectivity and permeability.

TABLE II

Permselectivities of the glassy polymers in Fig. 5, evaluated at 20 atm pressure of CO_2 and 20 atm pressure of CH_4 [53].

Polymer	$\frac{P_{CO_2}}{P_{CH_4}}$
Poly(phenylene oxide)	15
Polycarbonate	24
Polysulfone	28
Cellulose Acetate*	31
Kapton®	64

* Assumes equal % increase in CO_2 and CH_4 permeability due to any plasticization of cellulose acetate by CO_2.

TABLE III

Comparison of novel high-flux polymers with standard polysulfone (41, 42, 53, 54).

Polymer	P_{H_2}	$\frac{P_{H_2}}{P_{CH_4}}$	P_{CO_2}	$\frac{P_{CO_2}}{P_{CH_4}}$
A	32	427	—	—
B	76	112	—	—
Polysulfone	12	78	4.4	28
C	—	—	21	68
D	—	—	65	24

Permeability has units of Barrers
Temperature is 30–35°C
Note: The chemical structures of the polymers referred to in Table III are presented and discussed later (see Table VII).

These arguments can then be extended to define general directions for research to select interesting candidate polymers for advanced materials for the next generation of membrane processes.

The sorption isotherms for most gases in glassy polymers tend to have concave shapes like those shown in Fig. 6. As a result, the apparent solubility ($\bar{S} = C_2/p_2$) is a decreasing function of the upstream penetrant pressure.

On the other hand, the diffusion coefficients, $\bar{D}$, of gases in glassy polymers typically increase somewhat with sorbed concentration even in the absence of strong plasticization (43). Such increases are moderate for all of the materials in Fig. 5 except cellulose acetate and can be explained in terms of saturation of excess volume in the polymer as concentration increases (43). Strong plasticization, conversely, produces dramatic

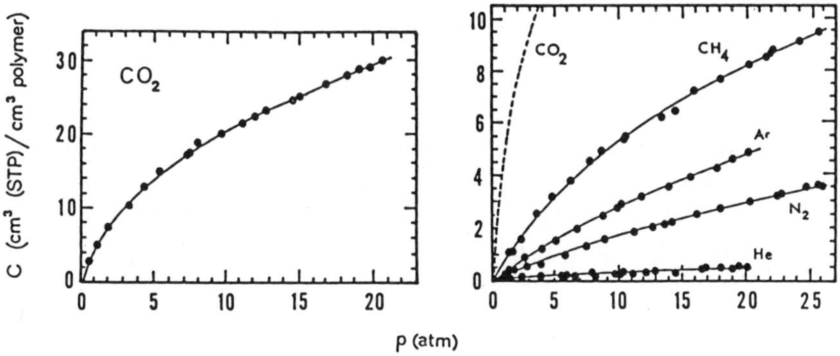

FIGURE 6 Sorption isotherms for various gases in polycarbonate at 35°C [55].

upswings in the apparent mobility, as appears to be the case with cellulose acetate. The observed pressure dependencies of the permeabilities shown in Fig. 5 are the net result of these two effects related to $\bar{D}$ and $\bar{S}$.

In the case of cellulose acetate, the sorption isotherm for CO_2 has the same type of concave shape as shown in Fig. 6, so that a sharp increase in $\bar{D}$ apparently swamps the effect of the decreasing apparent solubility coefficient. The reason for cellulose acetate's greater tendency to be plasticized compared to the other polymers in Fig. 5 may be the high concentration of carbonyl groups present in its structure (56). Since plasticization tends to produce a more rubber-like material, selectivity losses are expected ultimately to attend the large upswing in CO_2 permeability. Data by Wensley (35) suggest that serious losses in cellulose acetate selectivity are not observed immediately as a result of the initial plasticization, but above CO_2 partial pressures of 20 atm, the drop in selectivity is clear. All materials shown in Fig. 5 are also expected to show an eventual plasticization response similar to that of cellulose acetate, but it will occur at higher pressures because of a lower tendency of these materials to interact specifically with CO_2, as is discussed below.

Solubility Selectivity

If one uses the definition of $\bar{s}$ given in Eq. (5), the apparent solubility coefficient for a gas such as CO_2 or CH_4 in a glassy polymer will be a function of pressure and will approach a limiting value at higher pressures. The solubility in rubbery polymers and low-molecular-weight liquids, on the other hand, does not show this trend. Consider the solubility of CO_2 in glassy poly(methyl methacrylate) (PMMA), polyethylene (adjusted to

100% amorphous basis), butyl rubber, n-hexane, n-heptane, and 2,2,4 trimethyl pentane (56) shown in Fig. 7a. Presenting these data in the form of $\bar{S}$ in Fig. 7b, the value for CO_2 in PMMA is seen to approach, in an asymptotic fashion, a constant value which is *above* that of the nonpolar rubbery polymers but considerably *below* that of the nonpolar low-molecular-weight liquids. The excess solubility coefficient of CO_2 in the PMMA [above the asymptotic value of 0.944 cc(STP)/cc atm] can be interpreted as resulting from additional sorption capacity in the glassy polymer due to the nonequilibrium nature of the glassy state (57, 58). This asymptotic solubility coefficient value (called k_D) is typically taken to be representative of true polymer/gas interaction. The curvature in the isotherm, on the other hand, is believed to be largely a function of prior history effects related to the nonequilibrium nature of glassy polymers. This limiting value, k_D, is therefore analogous to the true solvent/gas interaction in low-molecular-weight solvent media and simple polymer/gas interactions in rubbery polymers. As such, it is referred to in all three

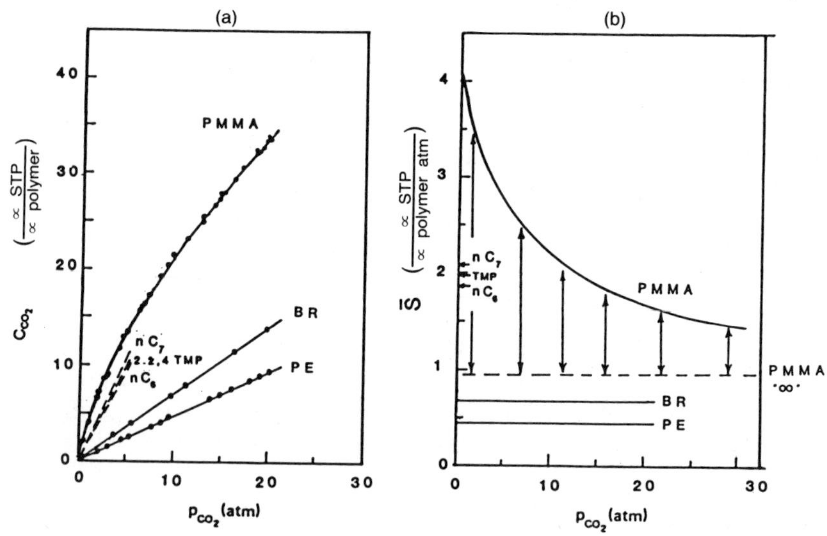

FIGURE 7 Isotherms (a) and solubility coefficients (b) for CO_2 in various media. PMMA = poly(methyl methacrylate), a typical glassy polymer. BR, PE = butyl rubber and polyethylene, respectively, typical rubbery polymers. The PE isotherm has been adjusted to a 100% amorphous basis. n-C_6, n-C_7, and 2,2,4 TMP = n-hexane, n-heptane, and 2,2,4 trimethyl pentane, respectively, are representative of typical low-molecular-weight solvent environments. The low-molecular-weight data are available only at 25°C and low pressure [56]. The polymer data were all measured in our laboratory at 35°C.

cases as the Henry's law (or more correctly the Bunsen) solubility coefficent for the gas in the respective media.

Comparison of the value of k_D for glassy PMMA to the corresponding solubility values for the nonpolar rubbery and liquid sorption media indicates several points. First, the sorption coefficient for the nonpolar rubbery polymers is lower than that for the polar polysulfone. Pilato *et al.* (42) have discussed this effect in terms of favorable interactions between the sulfone group in polysulfone and CO_2. Specific CO_2 interaction effects are also anticipated with the ester groups in PMMA and cellulose acetate (CA), as is discussed in more detail in a following section.

The higher sorption level of CO_2 in n-hexane compared to amorphous polyethylene is, however, clearly not due to differences in specific interaction of CO_2 with n-hexane and these rubbery hydrocarbon polymers. The higher solubility of penetrants in low-molecular-weight media of similar chemical structure has been interpreted in terms of two effects: a simple combinatorial entropy effect and a so-called free volume or equation-of-state effect (59–62).

The combinatorial effect arises primarily from the fact that solvent molecules are smaller than polymer molecules. If one assumes a similar molar volume for CO_2 in polymers and low-molecular-weight solvents, the entropy of mixing is obviously lower for CO_2 in high-molecular-weight polyethylene compared to hexane.

The simply Flory-Huggins equation accounts only for the pairwise contact enthalpic effects between solvent and solute and for the loss in configurational entropy of the polymer due to the connectedness of its subunits. Additional factors associated with inadequacies of lattice representations of fluids are not accounted for in this simple theory. These effects, addressed by Flory in his more sophisticated theory, are rather complex and depend upon the relative size of the solute and solvating molecule (59–62). Both enthalpic and entropic deviations from the simple Flory-Huggins representation of chemical potential can occur.

Fortunately, while the above effects related to combinatorial and equation-of-state differences between low-molecule-weight solvent environments and high polymer environments are interesting and appropriate for considering the *absolute* solubility of a component in different media, they become less important in considering the *ratio* of solubilities of two penetrants. Since these complicating effects act essentially as multiplicative factors upon the solubility coefficients of small penetrants of similar size (e.g., CO_2 and CH_4), they tend to cancel out upon taking ratios. This useful principle is illustrated well in Table IV, where the absolute value of solubilities and the ratios of solubilities for CO_2 and CH_4 are compared for various polymers and low molecular-

TABLE IV

Absolute solubility coefficients and solubility selectivities composed of ratios of solubility coefficients for CO_2 and CH_4 in a variety of media.

Medium	$[k_D]_{CO_2}$ $\left(\frac{\text{cc STP}}{\text{cc polymer atm}}\right)$	$[k_D]_{CH_4}$ $\left(\frac{\text{cc STP}}{\text{cc polymer atm}}\right)$	$\frac{[k_D]_{CO_2}}{[k_D]_{CH_4}}$
HDPE (100% amorphous basis)	0.453	0.225	2.01
Nat. Rubber	0.904	0.250	3.61
Poly(phenylene oxide)	0.95	0.33	2.88
Polysulfone	0.664	0.161	4.12
Polycarbonate	0.685	0.147	4.66
Kapton®	0.60	0.15	4.1
PMMA	0.94	0.10	9.26
Cellulose Acetate	0.849	0.075	11.39
n-Hexane	2.08	0.87	2.40
n-Heptane	1.83	0.53	3.47
Carbon Tetrachloride	2.42	0.66	3.66
Toluene	2.15	0.51	4.19
Benzene	2.48	0.53	4.68
Chlorobenzene	2.18	0.44	4.93
Ethanol	2.46	0.48	5.12
Methanol	3.54	0.48	7.36
Methyl Acetate	6.38	0.55	11.53
Acetone	5.82	0.54	10.75
Formamide	1.69		
DMSO	2.88	0.12	23.5
Water	0.77	0.03	25.8

weight solvents displaying a wide range of chemical characters (56, 63–65). Clearly, the solubility selectivities of similar low-molecular-weight components and polymers are similar even thought the absolute solubilities do not appear to agree. Compare, for example, high-density polyethylene (HDPE) and hexane, or cellulose acetate and methyl acetate.

The low-molecular-weight solvent data in Table IV indicate that as the polarity of the medium increases, the solubility selectivity for the CO_2/CH_4 system increases (63–65). Interestingly, the absolute solubility coefficient for CO_2 in polar and hydrogen-bonding media such as water, formamide, and DMSO is considerably lower than in less polar media such as acetone and methyl acetate. The partial molar volume of CO_2 in water has been found to be substantially smaller than in the other less associated media (64). It has been suggested that solvent association in the highly polar and hydrogen-bonding solvents "appear to have the effect of excluding solute molecules" (64). To optimize both the absolute solubility and selectivity of a liquid medium, a choice of acetone or methyl acetate would provide a reasonable compromise.

In the case of gas separation membranes, simultaneous maximization of absolute CO_2 solubility and solubility selectivity may not be desirable, since specific interactions between the penetrant and the polymer may promote plasticization at high pressures. Data are reported for various polymers in Table V (53) for overall separation factors as well as the mobility and solubility selectivities defined in Eq. (8). It is clear from comparing the data in Table V that cellulose acetate is superior to all other membrane materials considered in terms of solubility selectivity if one uses the ratios of k_D values for CO_2 and CH_4 for comparison. Solubility selectivities calculated from the ratio of $\bar{S}$ values for CO_2 and CH_4 evaluated at a partial pressure of 20 atm of each pure component are also reported in Table V. Similar, but less clear-cut differences among the various polymers are apparent if one uses the 20 atm $\bar{S}$ ratios reported in Table V instead of k_D ratios. The interpretation based on k_D values is preferable, since residual sorption contributions from excess volume tend to confound slightly the $\bar{s}$ values evaluated at 20 atm which have not toally reached their asymptotic values characteristic of simple polymer/penetrant interactions.

The impressive solubility selectivity of cellulose acetate polymers is not surprising when one considers the structure of cellulose acetate materials with a degree of substitution of 2.4–2.8. The large density of carbonyl groups in such a polymer imparts an environment somewhat similar to that of methyl acetate or acetone. As in the case of polyethylene, the

TABLE V

Permeability, diffusity, and solubility ratios for CO_2 and CH_4 in various glassy polymers at 35°C.

Polymer	$\frac{P_{CO_2}}{P_{CH_4}}$ *	$\frac{D_{CO_2}}{D_{CH_4}}$ *	$\frac{S_{CO_2}}{S_{CH_4}}$ *	$\frac{(k_D)_{CO_2}}{(k_D)_{CH_4}}$
Poly(phenylene oxide)	15.1	6.88	2.2	2.9
Polycarbonate	24.4	6.81	3.6	4.7
Polysulfone	28.3	8.85	3.2	4.1
Cellulose Acetate #	30.8	4.21	7.3	11.4
Kapton	63.6	15.38	4.1	4.1
PMMA				9.3

Assumes equal % increase in CO_2 and CH_4 permeability due to any plasticization of cellulose acetate by CO_2.

* Evaluated at 20 atm upstream pressure with negligible downstream pressure

apparent value of k_D for cellulose acetate is probably lower than the true value based only on the amorphous fraction. PMMA is a totally amorphous material with a high density of ester carbonyl groups like CA. PMMA has a higher CO_2 k_D value but a slightly lower CO_2-to-CH_4 solubility selectivity than cellulose acetate has.

Fig. 8 shows the relationship between solubility selectivity for the CO_2/CH_4 system and the mass density of carbonyl (or sulfone) groups for various polymers in Table V and for appropriate low-molecular-weight solvents (56). The abscissa values were calculated simply using the mass density of the various media and the weight fraction of the low-molecular-weight compound or polymer repeat unit that consists of the carbonyl (or sulfone) functionality. It is impressive that data for both low-molecular-weight and polymeric media can be correlated on this plot. Investigation of other carbonyl-containing low-molecular-weight compounds such as formaldehyde, acetaldehyde, etc., and polymeric media such as poly(methyl vinyl ketone) and poly(vinyl acetate) is desirable to test further the generality of the correlation. On the basis of Fig. 8 it is not surprising that the inherent solubility selectivity of polyphenylene oxide

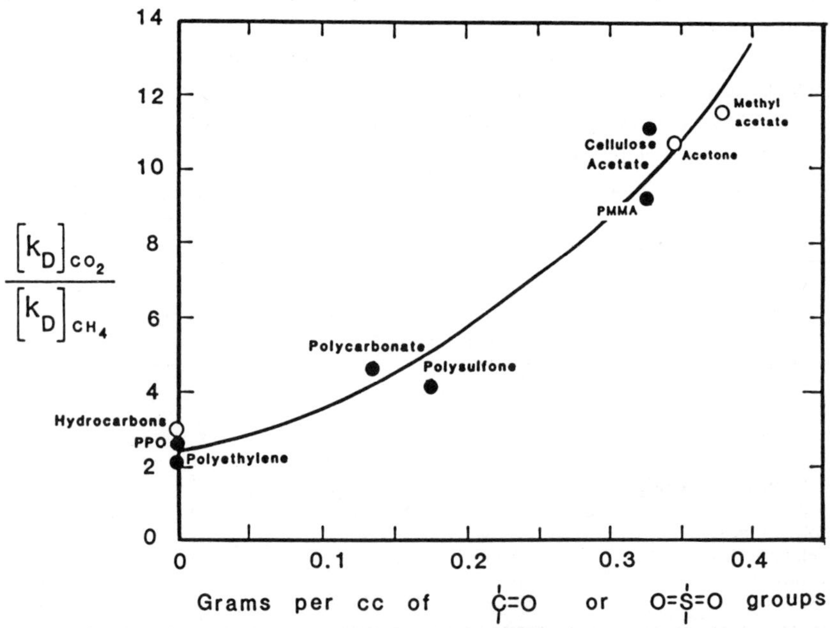

FIGURE 8 Correlation of the CO_2/CH_4 solubility selectivity of low- and high-molecular-weight compounds as a function of the mass density of carbonyl or sulfone groups in the medium.

(PPO), which lacks carbonyl groups, is only slightly higher than that of polyethylene for the CO_2/CH_4 system. The etheric linkage appears to be rather unimportant in specific interactions with CO_2.

It may ultimately be possible to formulate a more general abscissa coordinate to include other groups such as hydroxyls and amines with the ability to interact with CO_2, perhaps on the basis of acceptor-donor theory. Possible difficulties in implementing such a plan can be appreciated by considering the data for methanol and ethanol, which have densities of hydroxyl groups of 0.42 and 0.33 gm/cc, respectively. These can be converted to molar equivalents of 0.025 and 0.019 mole hydroxyl groups per cc, respectively, for comparison to the 0.014 and 0.012 mole carbonyl groups per cc, respectively, for methyl acetate and acetone. Clearly, hydroxyl groups are not as effective as carbonyl and sulfone groups for selective CO_2/CH_4 solubility enhancement, since even at higher molar equivalent concentrations, the alcohols in Table IV have only moderate selectivities. Differences between carbonyls and hydroxyls, also reflected in the absolute solubility values for CO_2 in the alcohols and esters, may be due to hydroxyl interaction with one another in preference to either CO_2 or CH_4. The solubility selectivity data for cellulose acetate in Table IV is for a degree of acetylation of 2.4/3.0, so there are still hydroxyl groups present. One might expect that the amorphous phase properties of cellulose acetate will be a strong function of its degree of acetylation, since this influences the ratio of carbonyl to hydroxyl groups.

The above solubility-enhancing considerations must, therefore, be buffered by the knowledge that as the tendency of a penetrant to interact with a membrane material is increased, a higher tendency to plasticize the material becomes likely.

Although the solubility selectivity of cellulose acetate is impressive, it is not large enough by itself to be competitive without a substantial complementary factor contributed by its mobility selectivity (Table V). If this mobility contribution is undermined by plasticization at high CO_2 pressure due to interactions that enhance the CO_2 solubility relative to that of CH_4, one might suspect that caution should be exercised in basing selectivity enhancement for future membranes on solubility effects.

As noted earlier, the form of the CO_2 permeability response for CA in Fig. 5 suggests that the selective CO_2/CH_4 interaction ability of CA indicated in Fig. 6 may, indeed, be associated with a greater tendency for CO_2 plasticization than is the case for the other polymers in Fig. 5. There are no reported data for high-pressure CO_2 permeability through PMMA, but on the basis of Fig. 8 and the preceding arguments, this polymer — like cellulose acetate — may also show a high tendency to be plasticized by CO_2. In a recent paper Wensley and Jakabhazy showed that the selectivity of CA tends to fall off more quickly compared to a corresponding polyimide under similar conditions (35). The structure of the polyimide is not reported, but it is noted that the material is a high glass transition polymer.

A related application in which solubility selectivity enhancement may be very useful is in the separation of air into its O_2-rich and N_2-rich components since the low solubility of O_2 in polymers makes it *unlikely* that plasticization would be a serious problem in this case.

Future Directions: Mobility Selection

The information currently available seems to suggest that concentrating on the "mobility selectivity" term in Eq. (8) is likely to be the most generally useful approach to guide the selection of polymers for advanced gas separation applications. Manipulation of the structure of standard bulk polymers to alter their mobility selectivity can be guided somewhat by correlations such as those in Fig. 9, which shows a correlation of $\bar{D}$ for various penetrants in a number of glassy polymers.

The correlating variable, the kinetic diameter of the gas, is related to the zeolite window dimension that will just permit passage of the penetrant of interest (66). For molecules that are essentially spherical, such as methane, argon, and helium, this dimension is very similar to the gas phase collision or Lennard-Jones diameter. For asymmetric molecules such as CO_2 and N_2, however, the kinetic diameter corresponds more closely to the minimum diameter of the molecule. This figure shows clearly the extremely strong effect on penetrant mobility caused by small differences in minimum effective penetrant diameter. Roughly 1.2 Å difference distinguishes the kinetic diameters of helium and methane; however, there are more than *three orders of magnitude* difference in their mobility in polycarbonate.

Fig. 9 is also useful for understanding the effects on permselectivity caused by variations in the polymer microenvironment resulting from variations in the nature of the membrane polymer. If one compares the relative magnitude of the diffusion coefficients of CH_4 in PPO and in Kapton®, for example, it is clear that there is roughly a five-hundred-fold higher CH_4 diffusion coefficient in PPO than in Kapton®. One may explain this observation in terms that are similar to those used to explain the observation of the thousand-fold difference in mobility between helium and methane in polycarbonate. Specifically, because of the difference in the backbone structures of PPO and Kapton®, the effective number of transient gaps of sizes 3.8 Å and larger that open near a CH_4

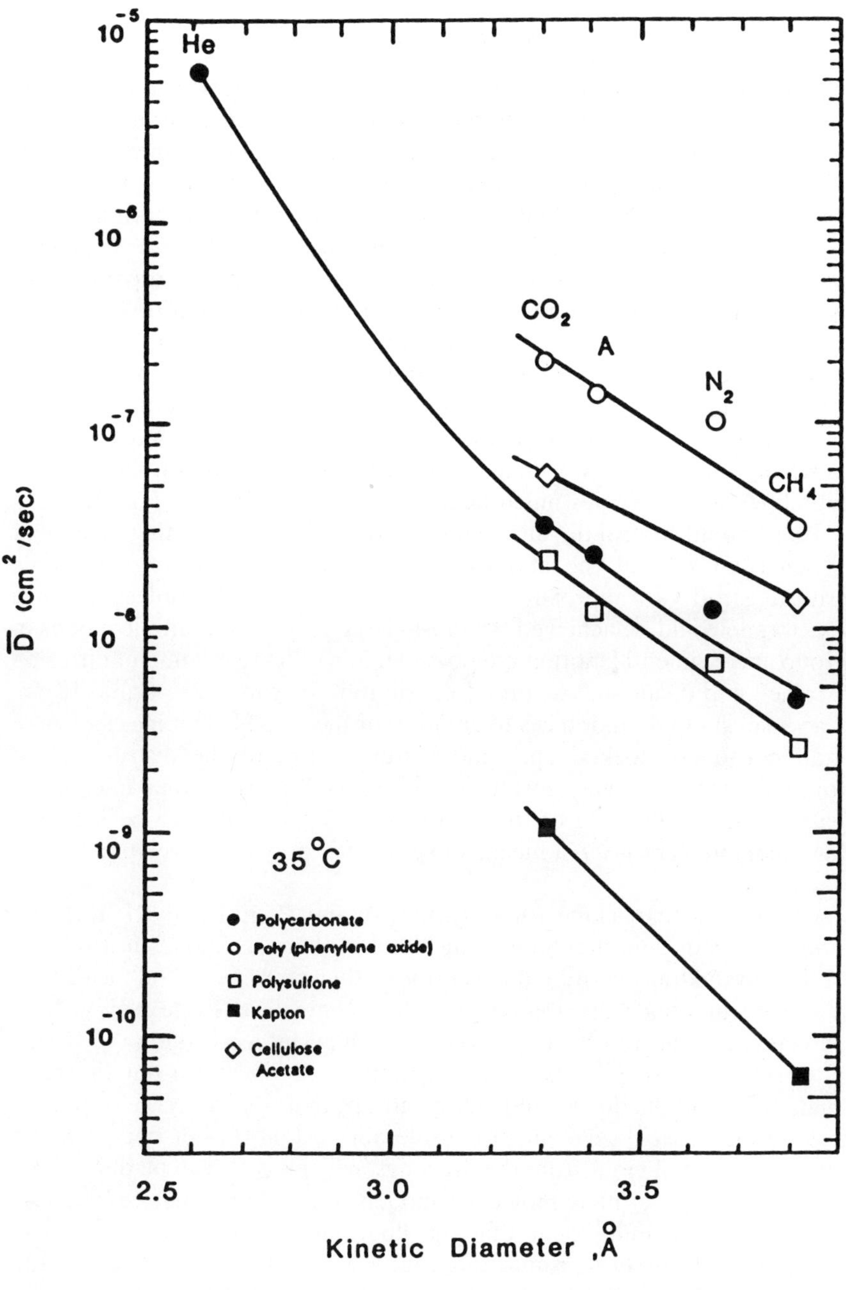

FIGURE 9 Correlation of the average diffusion coefficient, $\bar{D}$, and the kinetic diameters of several penetrants in a number of glassy polymers at 35°C for an upstream penetrant pressure of 20 atm.

penetrant is apparently roughly five-hundred times less in the mobility-restricted Kapton® than in the highly open PPO. The general tightening of the structure in Kapton®, however, causes a serious reduction in the diffusivity of *both* CO_2 and CH_4, thereby supporting the conventional wisdom that low permeability and high selectivity tend to correlate.

Table VI presents sieving dimensions for a number of important penetrants. It is clear that in addition to simple gases, separation of isomers such as n-C_4 and i-C_4 should be reasonably straightforward if one can avoid "over tightening" the structure of the membrane in order to achieve the necessary size discrimination. Indeed, a general but selective "opening up" of the membrane appears to be a wise direction to seek to provide reasonable productivities. If one resorts to standard glassy polymers such as PVC, their tight structure yields unattractively low mobilities for molecules much larger than CO_2 or O_2 (see Fig. 4).

If one could control the distribution of transient gap sizes that open near a penetrant so that the majority of such gaps are smaller than some critical cutoff, an approximate analog to a polymeric molecular sieve membrane could be achieved. In most glassy polymers, multiple modes of group and segmental motion are possible, and strict regulation of diffusion "traffic" is difficult since a broad distribution of gap sizes occurs. If one places substantial hindrances to rotation of the chain backbone, modes of motion can be "locked up," and better control of the distribution of transient gap sizes may result. If this procedure is accompanied by a purposeful modification of the chain to *inhibit* its ability to pack tightly (as occurs in Kapton®), a means of systematic polymer development is at hand.

Hindrances to packing for polymers A and B in Table III are clear when one considers the "kinks" included in their backbones in the form of the tetrahedrally coordinated central methylene carbons and the several non-colinear bonds distributed along the chain (see Table VII). These polymers are the result of research by Hoehn and coworkers (41, 67).

Moreover, strong hindrances to rotation are obvious from the rigid imide links and the fluorinated methyl groups that transform the bisphenol residue into a rigid segment of considerable extent. Other important information, not obvious from the drawings in Table VII, can be obtained if one uses the space-filling molecular models of these polymers (67). These models make it quite clear that the fluorinated methylene groups lose essentially all ability to rotate (an ability which they still possess in the unfluorinated material). Perhaps even more important is the fact that the fluorination of the methylenes allows the two phenyl groups to assume essentially only *one* permissible nonlinear conformation with respect to one another, thereby sculpting not only the length but also the shape of

TABLE VI

Kinetic (sieving) diameters of penetrants (66).

Molecule	He	H_2	NO	CO_2	A	O_2	N_2	CO	CH_4	C_2H_4	Xe	C_3H_8	n-C_4	CF_2Cl_2	C_3H_6	CF_4	i-C_4
Kinetic diameter (Å)	2.6	2.89	3.17	3.3	3.4	3.46	3.64	3.76	3.8	3.9	3.96	4.3	4.3	4.4	4.5	4.7	5.0

TABLE VII

Structure of novel polymers, discussed in Table III, that display both high permeabilities and selectivities.

POLYMER	STRUCTURE
A	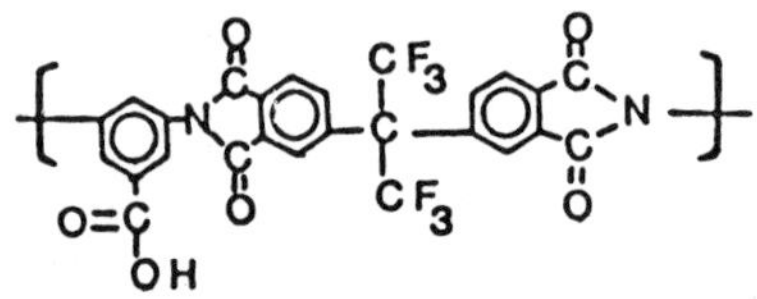
B	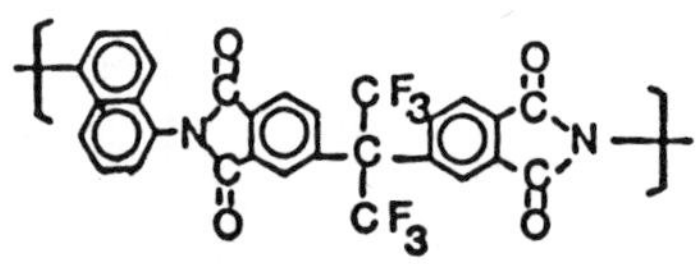
POLYSULFONE	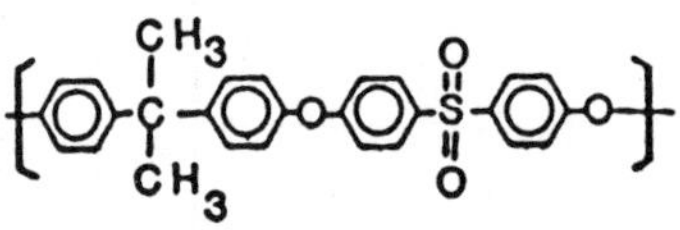
C	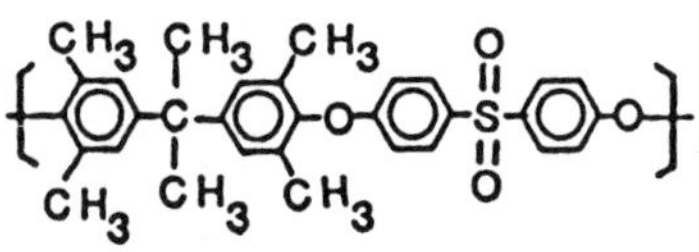
D	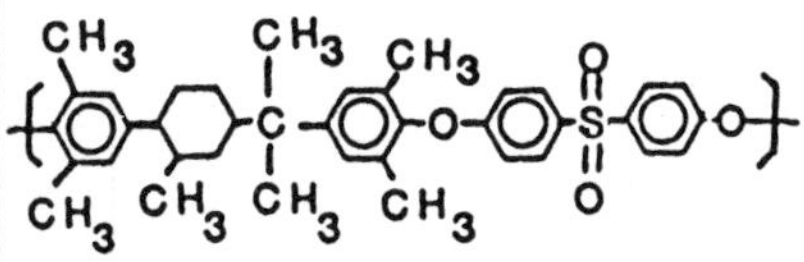

this rigid shape-regulating member. Although these materials are extremely interesting, monomers are not commercially available to permit their synthesis.

The polymers C and D, also shown in Tables III and VII, rely on differences in the substituent groups attached to the bisphenol-A residues to provide spacing and resistance to rotation. These polymers are the result of work by Pilato and coworkers (42). Consistent with the above discussion, the central methyl groups on the bisphenol-A residue can rotate.

Again from molecular model-building, one finds no strict constraint on the orientation of the phenyl groups in this residue, as in the case of the A and B fluorinated polymers, so that a distribution of such orientations is likely to exist. As noted in the context to polymers A and B, the central tetrahedral carbon provides a definite nonlinearity to the molecule, thereby inhibiting packing in both the substituted and unsubstituted materials.

An extremely interesting facet of the rigid fluorine-containing polymers is the report by Hoehn that their ultimate separating abilities are not developed fully until they have been annealed for roughly 6 hours at 260°C (67). In many cases, the permeability of the desired component was reported to increase slightly while the selectivity of the membrane increased substantially. It is likely that such annealing permits small rearrangements of chain segments as traces of excess volume are removed and redistributed. As the various segments "settle" into conditions more representative of their energetically desired state, it appears that the inherent kinks and rotational resistances come into full play. One may wonder, in fact, whether the lower selectivity of polymer D shown in Table VII might be elevated by annealing *without* a large drop in selectivity similar to observations by Hoehn in the patent dealing with fluorinated materials (67).

Plasticization

The data in Fig. 9 pertain to the situation in which the upstream pressure is 20 atm. In the absence of strong plasticizing (characterized by upswings in the permeability-versus-pressure curve such as seen for cellulose acetate in Fig. 5), the general trend in diffusivity values as a function of kinetic diameters is likely to be modified only slightly at different pressures. For example, Fig. 10 shows the range of values of $\bar{D}$ for penetrants measured in polycarbonate at 1 atm to compare to the 20 atm values discussed earlier. It is obvious that while the absolute values of the diffusion coefficients are different at the two pressures, the trends are similar, and one would draw conclusions from the 1 atm data that are consistent with those noted earlier in considering the 20 atm values.

It is, nevertheless, extremely important to explore pressure ranges even beyond 20 atm to determine the point at which plasticization begins to be a serious consideration, since plasticizing tends to make simple correlations — such as those shown in Figs. 9 and 10 — fail. For each polymer/gas system the point of serious plasticization must be determined experimentally. After identifying the partial pressure range for a given

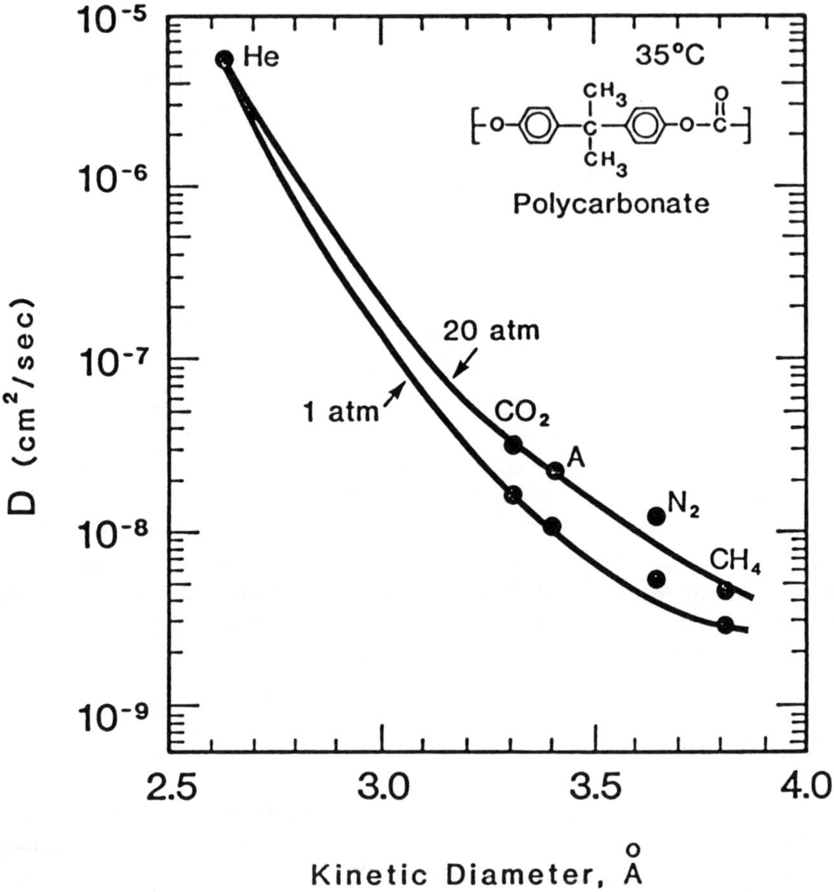

FIGURE 10 Representation of the degree of difference between effective diffusion coefficients measured at 1 atm and at 20 atm upstream pressure.

penetrant (e.g., CO_2), for which pure-component data suggest incipient plasticization, detailed multicomponent permeation experiments should be performed to characterize the abruptness of the actual selectivity loss following the initial upswing in permeability. Based on reports by Wensley, selectivity of CA is not lost significantly up to roughly 300 psia even though the data in Fig. 5 suggest the existence of plasticization. The samples studied by Wensley and reported on in Fig. 5 are different, so caution must be exercised; however, it may be that selectivity is not lost immediately when the upturn in permeability is observed. The high T_g

polymers in Table VII are expected to have extremely high inherent use temperatures and resistance to plasticization; however, they too will presumably ultimately exhibit plasticization-type behavior under demanding conditions of extremely high CO_2 pressure.

ACKNOWLEDGEMENTS

The author gratefully acknowledges the support of the National Science Foundation under grant number CPE-8319285 and of The University of Texas at Austin Center for Energy Studies.

REFERENCES

1. Gardner, R.J., Crane, R.A., and Hannan, J.F., *Chem. Eng. Prog. 73* (11), 76 (1977).
2. Bollinger, W.A., MacLean, D.L., and Narayan, R.S., *Chem. Eng. Prog. 78* (10), 27 (1982).
3. Schell, W.J., and Houston, C.D., *Chem. Eng. Prog. 78* (10), 33 (1982); *Hydrocarbon Process. 61* (9), 249 (1982); and paper #83C presented at AIChE Symposium, Houston, Tex., Mar. 28–31, 1983.
4. Mazur, W.H., and Chan, M.C., *Chem. Eng. Prog. 78* (10), 38 (1982).
5. Coady, A.B., and Davis, J.A., *Chem. Eng. Prog. 45* (10), 45 (1982).
6. Youn, K.C., Blytas, G.C., and Wall, H.H., paper given at Gas Processors Association Regional Meeting, Houston, Tex., Nov. 1983.
7. Burmaster, B.M., and Carter, D.C., paper #83A presented at AIChE Symposium, Houston, Tex., Mar. 28–31, 1983.
8. Schell, W.J., *Hydrocarbon Process. 62* (8), 43 (1983).
9. Russell, F.G., *Hydrocarbon Process. 62* (8), 55 (1983).
10. Lane, V.O., *Hydrocarbon Process. 62* (8), 56 (1983).
11. Schendel, R.L., Mariz, C.L., and Mak, J.Y., *Hydrocarbon Process. 62* (8), 58 (1983).
12. Shultz, J.S., Goddard, J.D., and Suchdeo, S.R., *AIChE J. 20*, 417 (1974).
13. MacLean, D.L., and Graham, T.E., *Chem. Eng. 87* (4), 54 (1980).
14. Lonsdale, H.K., *J. Membr. Sci. 10*, 81 (1982).
15. Ward, W.J., Browal, W.R., and Salemme, R.M., *J. Membr. Sci. 1*, 99 (1976).
16. Kelly, Susan C., news release, "Separation Technology Used for Atmospheric Blanketing," Monsanto Public Relations Dept., Oct. 1983.
17. Stannett, V.T., Chap. 2 in *Diffusion in Polymers*, J. Crank and G.S. Park (eds.), Academic Press, N.Y., 1968.
18. Bixler, H.J., and Sweeting, O.J., in *The Science and Technology of Polymer Films*, Vol. II, O.J. Sweeting (ed.), Wiley, N.Y. (1971), p. 2.
19. van Amerongen, G.J., *Rubber Chem. Technol. 37* (5), 1065 (1964).
20. Stannett, V.T., Koros, W.J., Paul, D.R., Lonsdale, H.K., and Baker, R.W., *Adv. Polym. Sci. 32*, 99 (1979).
21. Kammermeyer, K., *Ind. Eng. Chem. 49*, 1685 (1957).
22. Robb, W.L., *Ann. N.Y. Acad. Sci. 146*, 119 (1967).
23. Stannett, V.T., Koros, W.J., Paul, D.R., Lonsdale, H.K., and Baker, R.W., *Adv. Polym. Sci. 32*, 99 (1979).
24. Stern, S.A., "Gas Permeation Processes," in *Industrial Processing with Membranes*, Wiley-Interscience, N.Y. (1972).

25. Chern, R.T., Koros, W.J., Fedkiw, P.S., paper presented at the Winter National AIChE Meeting, Atlanta, Ga., 1984.
26. Antonson, G.R., Gardner, R.J., King, C.F., and Ko, D.Y., *Ind. Eng. Chem. Process Des. Dev. 16* (4), 463 (1977).
27. Gerow, J.A., US Patent 3,832,830, Sept. 3, 1973, E.I. duPont.
28. Chern, R.T., "Multicomponent Gas Permeation through Glassy Polymers: Measurement and Models," PhD dissertation, North Carolina State University, 1983.
29. MacLean, D.L., Prince, C.E., Chae, Y.C., *Chem. Eng. Prog. 76* (3), 98 (1980).
30. Shirley, J., Barzik, D., and Byrnes, D., *Chem. Process.*, Jan. 1982, p. 30.
31. Makin, E.C., and Okamoto, K.K., US Patent 4,181,675, Jan. 1, 1980, assigned to Monsanto Co.
32. King, W., Separex, Inc., personal communication.
33. Henis, J.M.S., and Tripodi, M.K., *Sep. Sci. Technol. 15* (4), 1059 (1980).
34. *Gas Processors Report*, J. Brewster (ed.), published by Spears Consulting Group, Oct. 10, 1983.
35. Wensley, C.G., and Jakabhazy, S.Z., "High Performance Gas Separation Membranes," paper presented at the Winter National AIChE Meeting in Atlanta, Ga., 1984.
36. Riley, R.L., Case, P.A., Lloyd, A.L., Milstead, C.E., and Tagami, M., paper presented at Joint Symposium on Water Filtration and Purification, AIChE and Filtration Society, Philadelphia, Pa., June 1980.
37. Parkinson, G., *Chem. Eng. 90* (11), 27 (1983).
38. Cadotte, J.E., King, R.S., Majerle, R.J., and Petersen, R.J., *J. Macromol. Sci. Chem. A15*, 725 (1981).
39. Crank, J., *The Mathematics of Diffusion*, 2nd ed., Oxford Press, Clarendon, (1975).
40. Berens, A.R., *J. Vinyl Technol. 1*, 8 (1979).
41. Pye, D.G., Hoehn, H.H., and Panar, M., *J. Appl. Polym. Sci. 20*, 287 (1976).
42. Pilato, L., Litz, L., Hargitay, B., Osborne, R.C., Farnham, A., Kawakami, J., Fritze, P., and McGrath, J., *ACS Reprints 16*, 42 (1975).
43. Chern, R.T., Koros, W.J., Sanders, E.S., Chen, S.H., and Hopfenberg, H.B., ACS Symposium Series No. 233, *Industrial Gas Separations*, T.E. Whyte, C.M. Yon, and E.H. Wagner (eds.), 1983.
44. Stern, S.A., and Walawender, W.P., *Sep. Sci. 4*, 129 (1969).
45. Rogers, C.E., in *Physics and Chemistry of the Organic Solid State*, D. Fox, M.M. Labes, and A. Weissberger (eds.), Interscience, N.Y. (1965).
46. Yi-Yan, N., Felder, R.M., and Koros, W.J., *J. Appl. Polym. Sci. 25*, 1755 (1980).
47. McCandless, F.P., *Ind. Eng. Chem. Process Des. Dev. 11*, 470 (1972).
48. Chern, R.T., Koros, W.J., Hopfenberg, H.B., and Stannett, V.T., *J. Polym. Sci., Polym. Phys. Ed. 22*, 1061 (1984).
49. Chern, R.T., Koros, W.J., Hopfenberg, H.B., and Stannett, V.T., *J. Polym. Sci., Polym. Phys. Ed. 21*, 753 (1983).
50. Chern, R.T., Koros, W.J., Sanders, E.S., and Yui, R.E., *J. Membr. Sci. 15*, 157 (1983).
51. Antonson, C.R., Gardner, R.J., King, C.F., and Ko, D.Y., *Ind. Eng. Chem. Process Des. Dev. 16*, 463 (1977).
52. Stern, S.A., Mauze, G.R., and Frisch, H.L., *J. Polym. Sci., Polym. Phys. Ed. 21*, 1275 (1983).
53. Chern, R.T., Koros, W.J., Hopfenberg, H.B., and Stannett, V.T., "Material Selection for Membrane-Based Gas Separation," accepted for publication in ACS Symposium, *Material Science Aspects of Synthetic Polymer Membranes*, D.R. Lloyd (ed.).
54. Koros, W.J., "Membrane-Based Gas Separations: Data Base and Models for Glassy Polymers," paper presented at Sunriver Membrane Conference, Redmond, Oreg., Sept. 1983.
55. Koros, W.J., Chan, A.H., and Paul, D.R., *J. Membr. Sci. 2*, 165 (1977).
56. Koros, W.J., "Simplified Analysis of Gas/Polymer Selective Solubility Behavior," paper presented at the Spring National AIChE Meeting in Anaheim, Calif., 1984.

57. Chern, R.T., Koros, W.J., Sanders, E.S., and Hopfenberg, H.B., *ACS Symposium Series, No. 223: Industrial Gas Separations*, T.E. Whyte, Jr., C.M. Yon, and E.H. Wagener, eds. (1983), p. 48.
58. Chern, R.T., Koros, W.J., Sanders, E.S., and Yui, R.E., *J. Membr. Sic. 15*, 157 (1983).
59. Flory, P.J., Orwoll, R.A., and Vrij, A., *J. Am. Chem. Soc. 86*, 3507 (1964).
60. Flory, P.J., Orwoll, R.A., and Vrij, A., *J. Am. Chem. Soc. 86*, 3515 (1964).
61. Flory, P.J., *J. Am. Chem. Soc. 87*, 1833 (1965).
62. Abe, A., and Flory, P.J., *J. Am. Chem. Soc. 87*, 1838 (1965).
63. Wilhelm, E., and Battino, R., *Chem. Rev. 73* (1), 1 (1973).
64. Hayduk, W., and Laudie, H., *AIChE J. 19* (6), 1233 (1973).
65. Horiuti, J., *Sci. Papers of the Inst. of Phy. and Chem. (Tokyo) 17*, 126 (1931).
66. Breck. D.W., *Zeolite Molecular Sieves*, Wiley, N.Y. (1974), p. 636.
67. Hoehn, H.H., US Patent No. 3,822,202, 1974.

Membrane Research and Applications in the U.S.

H. K. LONSDALE

Bend Research, Inc. Bend, Oregon

Spurred on by the successes of membrane processes in the past two decades, a good deal of membrane research is being conducted in the United States. The portion that is most accessible, of course, occurs in universities and research institutes. Among the important areas that can be cited are:

1. Gas separations research, particularly in the areas of new membrane materials, membrane fabrication techniques, and gas-processing methods.
2. Thin-film composite membranes, prepared by interfacial polymerization and other techniques.
3. Membrane reactors, for simultaneously performing chemical or biochemical reactions and effecting separation of products.
4. A variety of applications of membranes in the evolving biotechnology area.
5. Chemically resistant membranes, for use in ultrafiltration and other applications.
6. "Membrane electrodes," for use in quantitative analysis of species ranging from microions to biological macromolecules.
7. Liquid membranes, of both the immobilized and emulsion types, for carrying out a variety of separation and concentration applications.

This is a potpourri of my view of significant membrane research and applications in the United States. I want to talk about seven processes, four of which are new things not yet commercialized and some of which we are doing in our own lab. Figure 1 shows the three important pressure-driven water-separation processes: microfiltration, ultrafiltration, and reverse osmosis. Microfiltration and ultrafiltration do not really differ in the principles of the process. They are both basically molecular sieves of increasing fineness. Microfiltration is a process for removing particulates from water that are roughly one micron in size. Ultrafiltration filters particles another order or two of magnitude smaller. But it, too, is a sieving process. Reverse osmosis is fundamentally different. Reverse-osmosis membranes are not sieves, but basically solution-diffusion membranes. Things permeate reverse-osmosis membranes by dissolving in the membranes and diffusing through them.

PROCESS	CONCEPT	MATERIAL PERMEATING	FORCE	MATERIAL RETAINED
Micro-Filtration	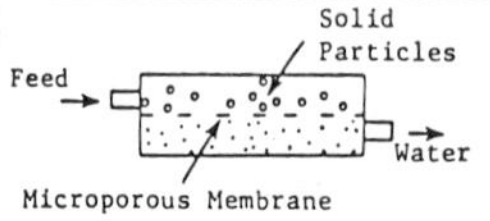	Water, Dissolved Solids	Pressure 10 psi	Suspended Silica, Bacteria >500,000 MW (0.01 μm)
Ultra-filtration	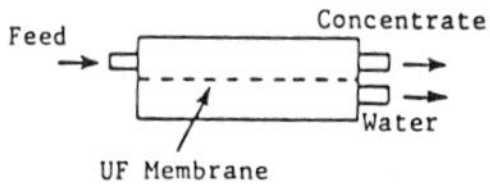	Water, Salts	Pressure 10-100 psi	Biologicals, Colloids, Macromolecules >5,000 MW
Reverse Osmosis	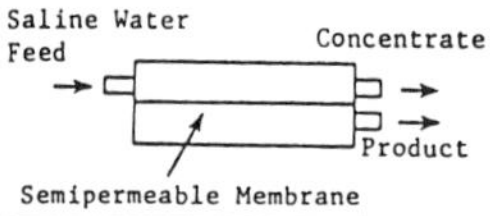	Water	Pressure 100-800 psi	All Dissolved, Suspended >50 MW

FIGURE 1

I am going to talk about some recent developments in ultrafiltration and in reverse osmosis. Figure 2 illustrates a crude break where microfiltration ends and ultrafiltration begins, and where ultrafiltration ends and reverse osmosis begins. These lines are obviously very fuzzy in terms of particle size differences and molecular weight cut-offs, but they are what most of the world considers to be the breaking points between those processes.

Figure 3 shows the non-pressure-driven liquid-separation processes. Michael Lysaght talked a little about the principal application of dialysis, which is purifying the blood of people with uremia using the artificial kidney. There are also some minor applications of dialysis, but I won't go into them any more than this. Gas permeation is a pressure-driven process for separating gases.

Figure 4 is a schematic of a chlor-alkali cell, which you heard something about from Dr. Leddy of Freeport. In this type of cell, sodium hydroxide and chlorine gas are made basically by the electrolysis of brines, and one needs a semi-permeable membrane to separate the two compartments. The new DuPont Nafion and the Japanese membranes work extremely well in this application. As you know, chlor-alkali cells with membranes are now replacing the mercury cells and asbestos cells that have been used for so many years.

Figure 5 is Lonsdale's version of the membrane market as of 1982. There are various ways of doing market guestimates. I have tried to include not just the cost of the membrane module, but also the cost of the equipment in which it is used. I would just point out that back when I was a kid, the worldwide membrane market was awfully small — less than $5 million a year — and now it is approaching a billion dollars a year.

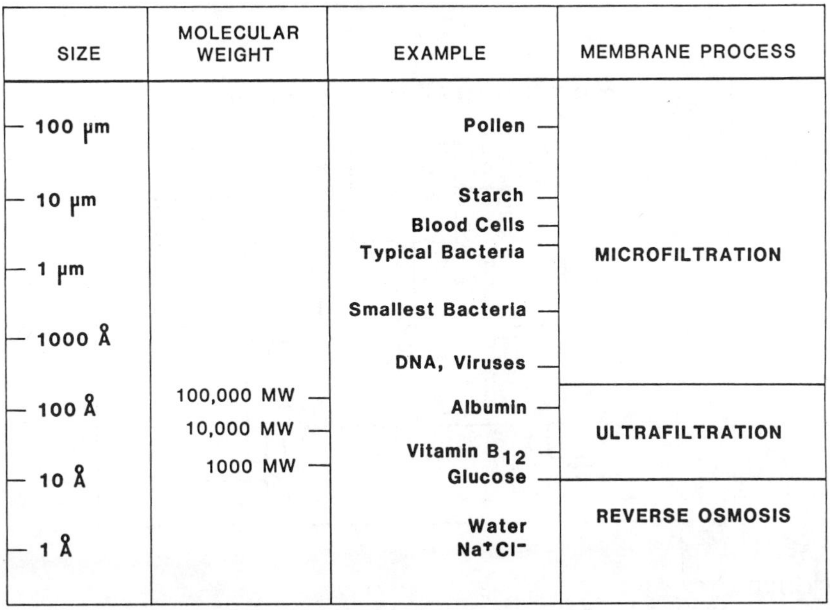

FIGURE 2

PROCESS	CONCEPT	MATERIAL PERMEATING	FORCE	MATERIAL RETAINED
Dialysis	Feed; Purified Stream; Inpurities; Dialysate Feed; Dialysis Membrane	Ions, Low MW Organics	Concentration	Higher MW Dissolved, Suspended >1,000 MW
Electro-dialysis	Concentrate; Product; Feed	Ions	Voltage	All Non-Iohic, Macromolecular
Gas Permeation	Feed; Lean Gas; Concentrated Gas; Semipermeable Membrane	More Permeable Gas	Pressure	Less Permeable Gas

FIGURE 3

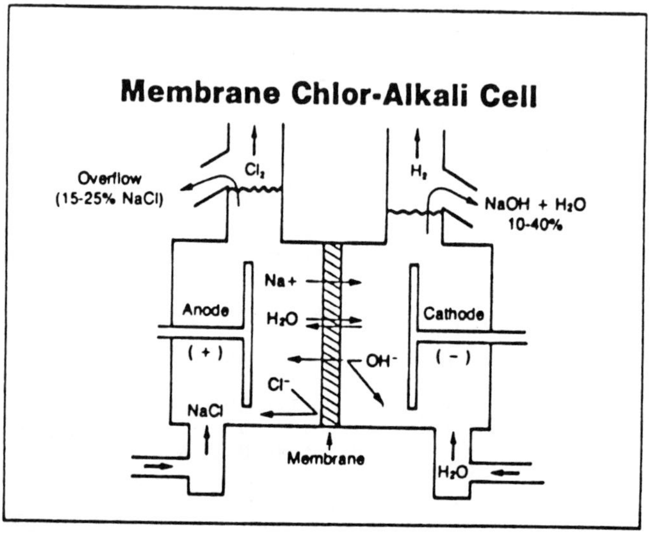

FIGURE 4

ESTIMATED WORLDWIDE MEMBRANE MARKET - 1982
Including Equipment

	$\$10^6$/YR
DIALYSIS, INCLUDING ARTIFICIAL KIDNEY: Early 1950's.	150
MICROFILTRATION: 1954.	250
ELECTRODIALYSIS: 1954.	60
CONTROLLED RELEASE: 1960's.	>100
REVERSE OSMOSIS: 1970.	120
MEMBRANE ELECTRODES: 1970's.	20
ULTRAFILTRATION: 1970's.	60
SEPARATORS FOR CHLOR-ALKALI CELLS, ETC.: Late 1970's.	30
GAS SEPARATIONS: 1980.	10
TOTAL:	$>\$800 \times 10^6$
PRE-1950 TOTAL:	$<\$5 \times 10^6$

FIGURE 5

Not all of the membrane market is in separation processes. Controlled release that Dave Enscore talked about yesterday is a major market — \$200 million a year is a bit of an understatement. The figure is more like \$200 to \$300 million a year. If you count skin patches, dog and cat flea collars, the tiny time capsules you hear about on TV — then controlled release in all of its manifestations is a several-hundred-million-dollar-a-year market. Another nonseparation process is membrane electrodes — still pretty small. I'll be talking about that process in some detail later on.

The membrane market is growing reasonably rapidly. It is not like microelectronics, which had an enormous and rapid growth, but it has been a fast-growing industry of its own. We see further important applications coming along, some of which we are working on, some of which we are not.

With that background, I'd like to turn to the first topic of my talk — gas separations. The principle breakthrough in that area is the Monsanto polysulfone hollow-fiber Prism^{TM} system introduced commercially about 4 years ago. Monsanto installed its first plants in captive, Monsanto sites about 6 years ago, and it has built quite a successful business for them — how large a business, I am not certain, but it certainly works. What they did, technically, to make it work was to take a membrane with a fairly high gas selectivity — polysulfone, specifically — and make an asymmetric membrane of it, which is fairly easy to do. In their early days those membranes tended to leak. You can't afford to have any leaks, particularly in gas permeation, because gas is such a fluid commodity that all of it will go right through the leak. You have to plug the leaks, and there have been various approaches to doing that. For example, you can take a piece of plastic tape and use a light box to find all the holes (people still do that). What Monsanto did, however, was very clever. They overcoated their polysulfone asymmetric membrane with a highly permeable second membrane — silicone rubber — that plugged the leaks and created a skin over the whole fiber. The skin doesn't interfere with the permeability of the membrane because silicone rubber is so permeable and so nonselective that all it does is plug the leaks. The flux can't go through it, yet the selectivity is improved enormously by this thin silicone coating. Figure 6 is a photograph borrowed from Monsanto that compares the size of their fibers with the size of a paper clip. The fibers are a few hundred microns in inside diameter and almost a millimeter in outside diameter. You can't see it, but the real fibers they use in their modules do have a thin silicone rubber coating no more than one micron thick. Silicone coating was an important breakthrough in membrane separations and Monsanto has capitalized on it. They have built several plants around the world — three or four dozen by now. Perhaps the most important application they have

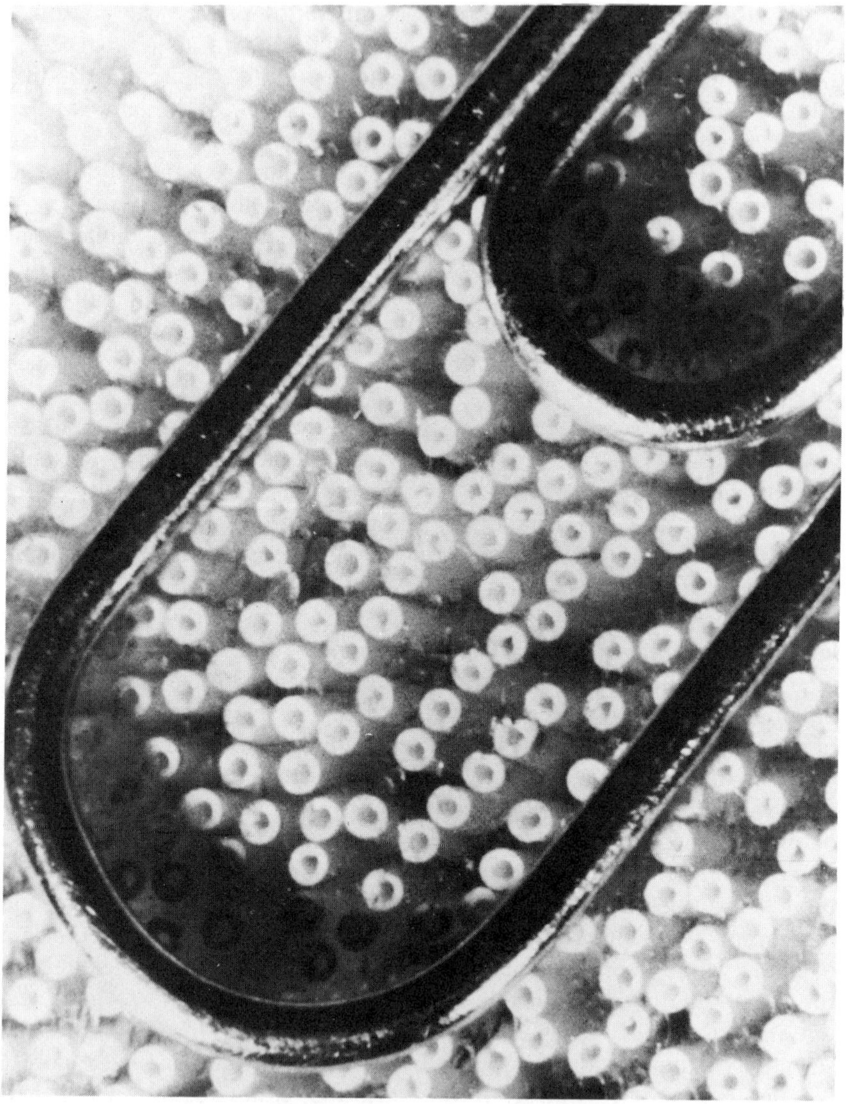

FIGURE 6

identified is recovery of hydrogen from ammonia synthesis. They take a reject stream from a reactor where ammonia is made, which contains more hydrogen than you would want to throw away, and put it through a hollow-fiber permeator like the one shown in Figure 7. The permeate can be released at either 400 psi or 1000 psi, or in fact at any pressure they might want to use. This permeate contains a highly enriched hydrogen stream that they can then feed back to the reactor and reuse. The nonpermeating stream, which is very low in hydrogen and has essentially no value, can be flared. Thus, from this feed, which has minimal value, can be recovered something of value. Hydrogen recovery and related applications have been quite good for Monsanto. Figure 8 shows their Louing, Louisiana plant at an ammonia facility. I think it is the first one they installed back

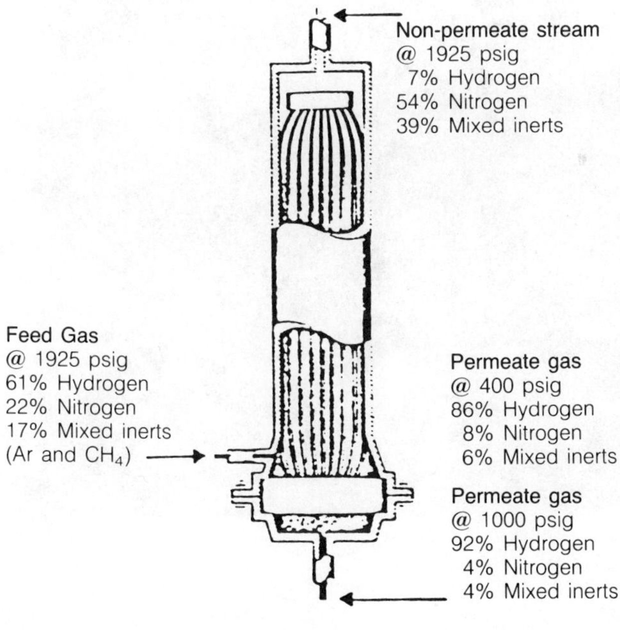

FIGURE 7

FIGURE 8

in 1978. The list of plants, shown in Figure 9, is now quite long. There are 33 plants shown here as of early 1983. They are all over the world: Brazil, Japan, China, The Netherlands, etc. Monsanto has been quite successful in introducing this new product, mostly for hydrogen recovery from ammonia streams. But as you can see from the list, they do several other types of separations, including water removal.

We at Bend Research have a joint venture with W.R. Grace Company to manufacture modules for Grace markets. Figure 10 shows our modules

INSTALLATIONS OF PRISM™ SEPARATORS AS OF 2/28/83

INSTALLATION/ LOCATION	APPLICATION	FEED VOLUME MM SCFD	START-UP DATE
1. Monsanto/Texas	H_2/CO Ratio Adjustment	4.5	March, '77
2. Monsanto/Florida	Hydrogenator Purge	1.0	April, '78
3. Monsanto/Texas	H_2/CO Ratio Adjustment	4.5	July, '79
4. Monsanto/Louisiana	Ammonia Purge	3.2	September, '79
5. California	Hydrodesulfurizer Purge	10.0	June, '80
6. Louisiana	CO Enrichment	1.5	September, '80
7. Monsanto/Alabama	Hydrogenator Purge	1.0	November, '80
8. Mississippi	Ammonia Purge	8.5	March, '81
9. Louisiana	Ammonia Purge	5.6	March, '81
10. Louisiana	Ammonia Purge	5.9	September, '81
11. Texas	Ammonia Purge	14.6	September, '81
12. Illinois	Ammonia Purge	6.1	October, '81
13. Germany	Ammonia Purge	6.3	November, '81
14. Japan	Adsorber Purge	.2	November, '81
15. Canada	Ammonia Purge	11.4	December, '81
16. Monsanto/Louisiana	Ammonia Purge	8.8	March, '82
17. Monsanto/Texas	Methanol Purge	5.2	April, '82
18. Mississippi	Ammonia Purge	11.8	June, '82
19. Italy	Gasoil Hydrofiner Purge	9.0	July, '82
20. Louisiana	Ammonia Purge	13.2	November, '82
21. Japan	Petrochemical	16.9	November, '82
22. Finland	Petrochemical Synthesis	7.4	November, '82
23. Louisiana	Ammonia Purge	5.6	December, '82
24. Alabama	Ammonia Purge	2.5	December, '82
25. Iowa	Ammonia Purge	5.6	February, '83
26. Netherlands	Ammonia Purge	6.2	2nd Quarter, '83
27. China	Ammonia Purge	8.3	2nd Quarter, '83
28. Pennsylvania	H_2/CO Ratio Adjustment	1.4	2nd Quarter, '83
29. Japan	Hydrodesulfurizer Purge	3.8	3rd Quarter, '83
30. Australia	Ammonia Purge	5.9	3rd Quarter, '83
31. Australia	Ammonia Purge	4.9	3rd Quarter, '83
32. Belgium	Petrochemical Purge	.5	3rd Quarter, '83
33. Brazil	H_2/CO Ratio Adjustment	2.9	1st Quarter, '84

FIGURE 9

in a unit installed in April of 1984 in Wyoming at an 8000-foot elevation — a pretty tough place to carry out an experiment. This six-pressure-vessel system removed most of the carbon dioxide from the high-CO_2 natural gas in a volume of about 3-1/2 million cubic feet a day. It is, to our knowledge, the largest gas test of its kind thus far in the world. We have another four or five tests like this going on in Texas and Colorado that are all working pretty well, so we expect this system to be introduced on a substantial scale by Grace sometime late in 1985. Like I said, this system is for CO_2 removal. Many natural gas streams in the southwestern U.S. and in Canada are acidic — that is, they have CO_2 and H_2S in them. For various reasons one does not want to pipeline those gases. The high CO_2 and/or hydrogen levels reduce the heating capability of the gas from the standard 1000 Btu per cubic foot to some lower values that make the

FIGURE 10

gas unusable in standard home burners. To upgrade the heating value, or because the CO_2 is corrosive, one wants to remove it. In the case of hydrogen sulfide, it is poisonous. So, for several reasons, one wants to purify those gases, and a good way to do that is with permselective membranes. We developed a membrane for that application for Grace and it is being tested now on a number of applications — CO_2 removal and water removal being the two principal ones.

I want to turn to a whole new area of research taking place in the U.S. and elsewhere — thin-film-membrane research. What started this area of research was the Loeb-Sourirajan reverse-osmosis membrane developed back in the late 1950's. I am sure you have all heard the story several times, but these two then-graduate students at UCLA found a way of making skinned membranes. Figure 11 is a scanning electron micrograph of a cross-section of such a membrane. There is a two-micron mark to show that the thin, white area at the top of the membrane is something less than a micron thick. The rest of the membrane is just a highly porous, highly permeable support for that thin skin. For a long time, the only way to make skinned membranes without imperfections in them was by the Loeb-Sourirajan method. Beginning in the '60s and continuing into the '70s, the group at North Star Research and the group at General

FIGURE 11

Atomic looked for, and found, a way to make very, very thin imperfection-free membranes. The most important, I believe, was the thin-film-composite membrane approach, particularly involving the use of interfacial polymerization. Interfacial polymerization seems to be the wave of the future for solid membranes. I want to talk about it before I talk about liquid membranes, which I think are also promising. For solid membranes, these thin-film composites, or TFC membranes, look extremely good. You can make these membranes with an effective thickness of only a few hundred angstroms, without pinholes, and to some extent, tailor the chemistry for either liquid or gas separation. They are not now used for gas separation, but they could be one day. Figure 12 shows schematically what a composite membrane looks like. The substrate of such a membrane, typically with some reinforcing web in it, has had placed on top of it by one means or another a very thin, semipermeable film. The most important method of putting that film down is by interfacial polymerization, a process described almost 10 years go by people at North Star Research. Among the authors are Cadotte and Petersen. The way they did this, and the way it is still done, is to take a microporous supporting membrane — typically microporous polysulfone which is fairly easily fabricated — and wet it with an appropriate solution containing a

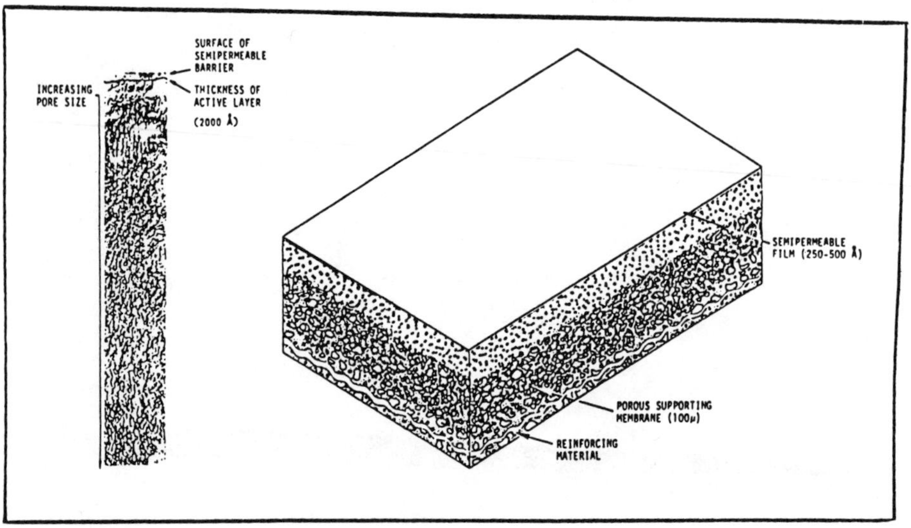

FIGURE 12

polymer or a prepolymer such as PEI. The membrane illustrated in Figure 13 is now known by the trade name NS-100; the NS stands for North Star Research Institute (no longer in existence), where this development was made. The membrane is wetted with the PEI, and one surface is contacted with a hexane solution of TDI. A reactive zone is created, which is subsequently cured at temperatures of 100°C to 150°C. This makes a crosslinked, fairly dense, very thin and imperfection-free film on one surface of the membrane. You also tend to get some "clutter," or some unreacted polymer on the substrate of the membrane. Although this can be a nuisance in some applications, one does get a very useful membrane with this method. The membrane is thin, has a high flux, and is without holes. There have been a number of variations played on this theme over the past 10 years in our laboratory. One superb water-desalting

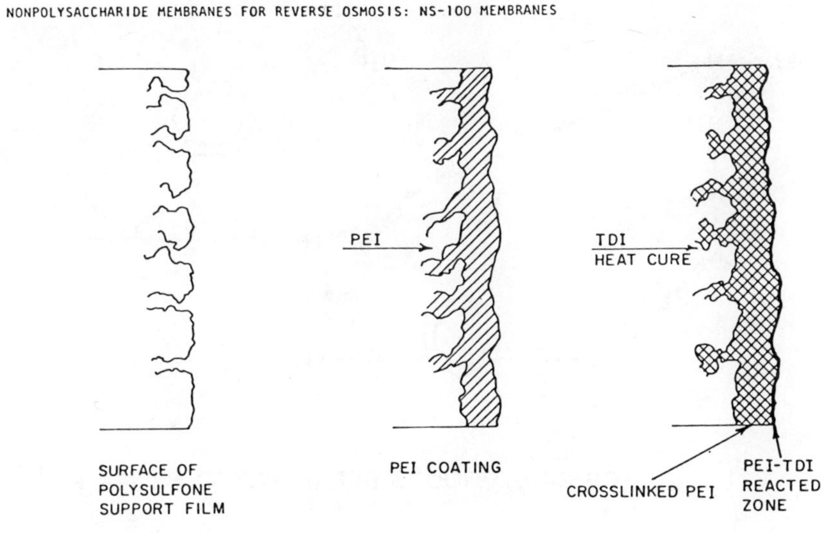

FIGURE 13 Schematic representation of the NS-100 membrane. From: L.T. Rozelle, *et al.*, "Nonpolysaccharide Membranes for Reverse Osmosis: NS-100 Membranes" in *Reverse Osmosis and Synthetic Membranes*, S. Sourirajan (Ed.), National Research Council of Canada, Ottawa, 1977.

membrane known by the trade name FT-30 is sold around the world by FilmTec and their licensees, and is perhaps the finest reverse-osmosis desalting membrane in the world today. This chemistry, and other chemistries such as the polyamide chemistry, remind all of us of the work that Paul Morgan did at DuPont 25 or 30 years ago that developed into what is known now as the "nylon rope trick." Interfacial polymerization is a very powerful technique for making high-flux, imperfection-free membranes. The chemistry of the NS-100 membrane is shown in Figure 14. Essentially, you get a three-dimensional, very dense, highly cross-

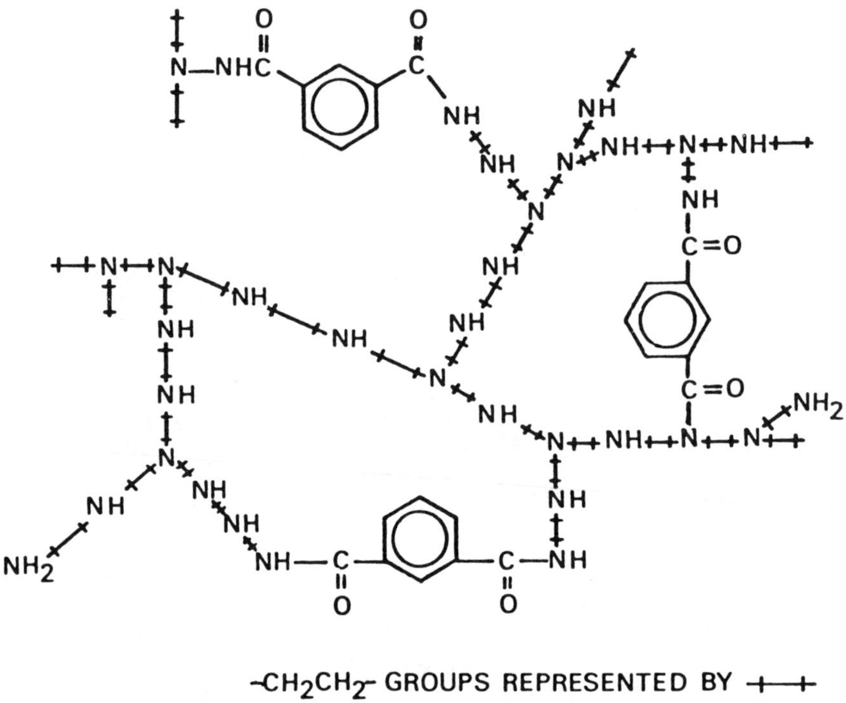

FIGURE 14 Chemical Structures of a Reverse-Osmosis Membrane Formed by Interfacial Polymerization.

linked product. Density is the key to making highly selective membranes. Rubbery polymers tend not to yield the selectivity one wants.

I want to turn now to something we refer to in our company as a membrane reactor. This is quite new. Membrane reactors come in as many sizes and shapes as does the membrane reactor concept. The membrane reactor concept involves combining a permselective membrane with some sort of reaction. There have been several variations of membrane reactors developed in laboratories around the world. I am going to talk about only one of those variations, which was developed in this country by Steve Matson as part of his Ph.D. thesis at the University of Pennsylvania. The variation was first taken to General Electric, then to our laboratory, and is now spreading to other places. The Matson Membrane Reactor (and I call it that because there are other versions) looks like the schematic in Figure 15. It is a composite membrane, one layer of which is an immobilized-liquid membrane, ILM for short, with a hydrophobic liquid imbibed into the pores of the cell by capillary action. I will tell you what this layer does in a minute. The second layer of the membrane is an immobilized enzyme. In our case, and in the work I am going to describe in the next couple of minutes, the enzyme chymotrypsin was covalently bound to the membrane, specifically to a microporous support, which is easy to do. Putting the two layers together is also easy to do, so making this membrane is not very hard. The membrane reactor application chose to study in a program that just ended in our company was the optical resolution of a racemic mixture, particularly the mixture of D- and L-phenylalanine: acetyl phenylalanine methyl ester (APME). What happens with this membrane is what both the D and the L forms of the ester can freely permeate this immobilized-liquid membrane. Because the immobilized liquid is hydrophobic, there is no reason both forms can't permeate. So, the esters diffuse across the membrane. As a matter of fact, the D-ester will also diffuse across the aqueous, enzyme-

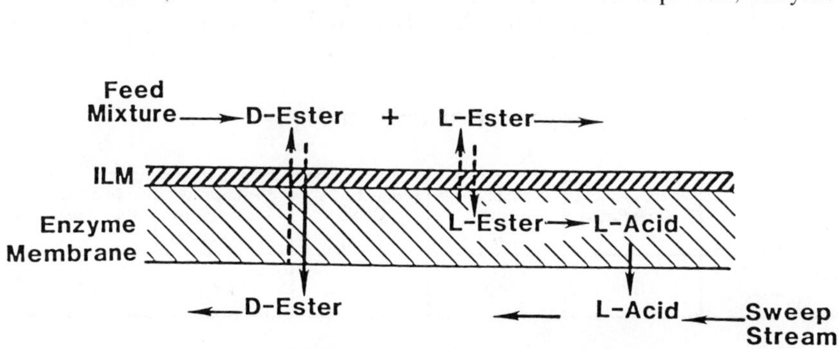

FIGURE 15

containing membrane. The aqueous membrane is porous; the pores are filled with water and lined with chymotrypsin. The D-ester can permeate quite freely and it winds up on the downstream side of the membrane, which is the product solution. The L-ester is specfically reacted with the chymotrypsin, which cleaves the ester to form the L-acid. Because the sweepstream is kept at neutral or basic pH, the L-acid is in an ionized form. The immobilized-liquid membrane is impermeable to the back-flow of that L-acid. The L-acid can only go one way; it can't go back. I emphasize that because you can play a cute trick with the system when that is the case. If you flow one stream at one rate and the other stream at the much lower rate, you can actually concentrate the L-acid in this stream; it simply has to go downward, through the membrane, and if you flow the two streams at different rates, for example a difference of a factor of 10, you can concentrate this L-acid by almost a factor of 10 — not quite 10, but still quite good. This membrane reactor then does two things: 1) it converts the L-ester into the L-acid, and 2) the L-acid ends up in the product stream along with the D-ester. But the separation of D-ester and L-acid is very easy to make by solvent extraction, and that's what we did. You shake this product up with a chloroform/water mixture or hexane/water mixture, and all the ester winds up in the organic phase and all of the acid ends up in the water phase. That is a very easy separation to make and you have a reasonably optically pure L-acetyl phenylalanine. You have taken the racemic mixture — a mixture that is essentially impossible to separate chemically — and converted it to something you can separate very easily with a membrane reactor. The other thing you have done is concentrate the L-acid by as much as you want to by adjusting the ratio of the two flow rates. Figure 16 shows the feed concentration in millimoles of the D-, L-acetyl phenylalanine methyl ester. The flow rate ratio of feed to product was about 1:30. The product concentration of L-acetyl phenylalanine was increased from about four millimoles to about 40 millimoles. The product APME concentration,

MEMBRANE REACTOR RESULTS

Feed D,L-APME Conc. (mM)	Feed: Product Flow Ratio	Product L-AP Conc. (mM)	Product APME Conc. (mM)	Product AP Purity (% of total solids)	Product Enrichment
3:8	32	40	2.8	94	21

FIGURE 16

which includes D-APME, is equal to the feed concentration, meaning that D-APME permeated the membrane freely and some of the L-APME did, too. The product purity of the acetyl phenylalanine, the desired product, which is presumably only L-AP (we checked it later by polarimetry) is almost completely pure — not quite, but awfully close. The optical purity is about 96%, meaning we had almost no D-AP in there. The product enrichment, in keeping with what I explained earlier, is comparable to the feed-to-product flow rate. In this case we were able to concentrate it about 20/fold. We have a long way to go yet to make aspartame — and you realize that was the reason we were looking at phenylalanine, not because it is a pretty rose, but because phenylalanine is the precursor to making the synthetic sweetener aspartame. Aspartame is made by reacting the phenylalanine methyl ester with L-aspartic acid, the L-isomer of phenylalanine methyl ester. We do not make that; we make acetyl phenylalanine. One could subsequently make phenylalanine methyl ester. We want to look at that one too. It might take a different enzyme or a somewhat different chemistry, but that is where we are headed with membrane reactors. There are several things you could do with membrane reactors; optical resolution of isomers is only one. It is a powerful technique. And although the economics are far from proven at this point, they look pretty good.

Biotechnology is another use of membrane reactors. This was worked on a bit at Stanford during the past couple of years and led to one facet of a spin-off company called Engenics in the Palo Alto area, established by Channing Robertson, Alan Michaels, and others. Figure 17 shows their version of a membrane reactor. It is a hollow fiber cut in half. It is a nice twist and could find real application. There is a 300-micron fiber with a feed stream containing some substance that you want to convert enzymatically to a useful product. Down the lumen of that fiber, immobilized one way or another in the pores of the wall of the fiber, is an enzyme. The substrate interacts with the enzyme by simply penetrating the fiber walls, thereby forming a useful product that ends up in the product stream. If you create just the right, slightly negative differential pressure across that fiber wall, you can get most of your product to go off that way. This idea is not yet commercial; I do not even know what its applications are likely to be, but it is a cute trick. The application of all kinds of membranes to biotechnology is gathering some momentum. I do not know of any important commercial applications yet, other than simple separations by microfiltration and ultrafiltration.

I want to turn now to another quite major departure — ultrafiltration. What has happened in ultrafiltration membranes over the last 25 years? There has been a fair amount of good work done there, I believe. Figure

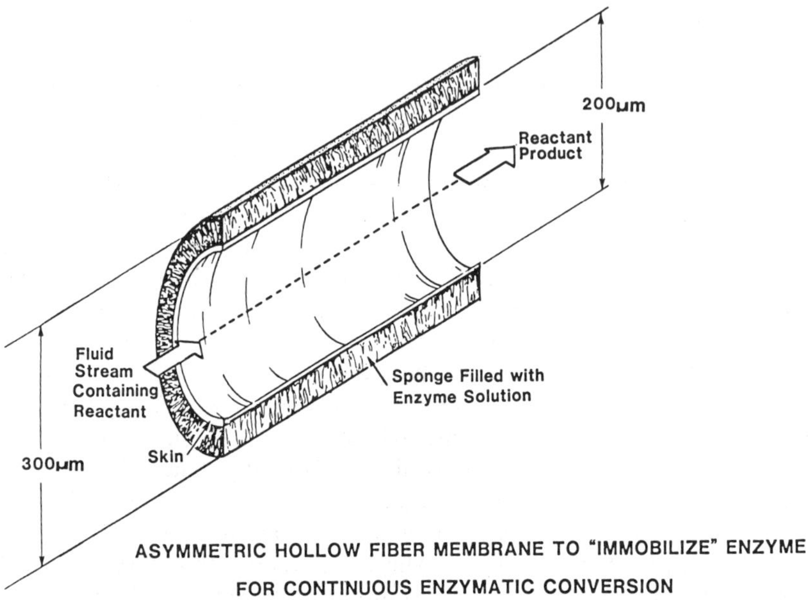

FIGURE 17

18 summarizes it. In the early days of ultrafiltration, people used a Loeb-Sourirajan-type unannealed, cellulose acetate membrane. You can vary the permeability and molecular-weight cutoff of these ultrafiltration cellulose acetate membranes by playing games with the casting solution and the casting conditions. For the first several years of UF, the only real problems with cellulose acetate were that you could not dry it out very well and it is not chemically as stable as some of the other polymers. What I have tried to show in this figure is the evolution of ultrafiltration membranes. People still use cellulose acetate in places, but there are now several other choices available. Dynel® was developed by Amicon in the early days. There was a polyelectrolyte complex — I don't think it ever became an important commercial success — but it was also developed by Amicon. Polysulfone came along 15 years ago and it has been an important commercial success, now used by several companies in the U.S. and elsewhere. Polyamides are important in and around the UF world. Abcor has developed polyvinylidene fluoride, which is a Kynar-type blend membrane with pyromellitic acid, and has done reasonably well. The Nitto Company in Japan marketed a crosslinked polyvinyl alcohol UF membrane. Polyvinylidene fluoride is obviously a very chemically resistant polymer, insoluble in almost all solvents — not quite

ULTRAFILTRATION MEMBRANES

- **Cellulose Acetate**
- **PAN/PVC Copolymer**
- **Polyelectrolyte Complex**
- **Polysulfone**
- **Polyamides**
- **Polyvinylidene Fluoride/ PMMA Blend**
- **Crosslinked Polyvinyl Alcohol**
- **Polyvinylidene Fluoride**
- **Polyimides**

FIGURE 18

insoluble in all of them, but it has now been made into a fairly useful membrane. I don't know if it is commercially sold, but it's awfully close right now. It is no Teflon yet, which is probably the ultimate. What people are working for, of course, are UF membranes that are chemically very stable — membranes that you could clean with acid, base, strong chlorines, solvents, etc. Polyimides are also chemically very stable. What we've gone for is new UF membranes that exhibit very high stability toward solvents, high temperatures, and other conditions of normal operation.

I want to do another shift of gears and talk about membrane electrodes. Membrane electrodes are not a large market, but they have an important future because people have found ways of making electrodes that are very specific for a large number of species. There are two very important approaches that involve membranes. These aren't your garden variety membranes, but if you extend the definition of membrane broadly enough, membrane electrodes fall into that category. What you need is something specific for one of the species in a solution — for example, a silver or silver chloride billet that could be formed into a membrane if you wanted to analyze for silver or chloride. One way or the other you have to have a reversible reaction occur at the membrane/solution interface that will create a very tiny ionic flow through the membrane crystal, which can be then converted into, or amplified as, an electrical signal. One approach is to use membrane crystals as illustrated in Figure 19 and another is to use liquid membranes as shown in Figure 20. The porous membrane shown in Figure 20, which contains an immobilized liquid, can

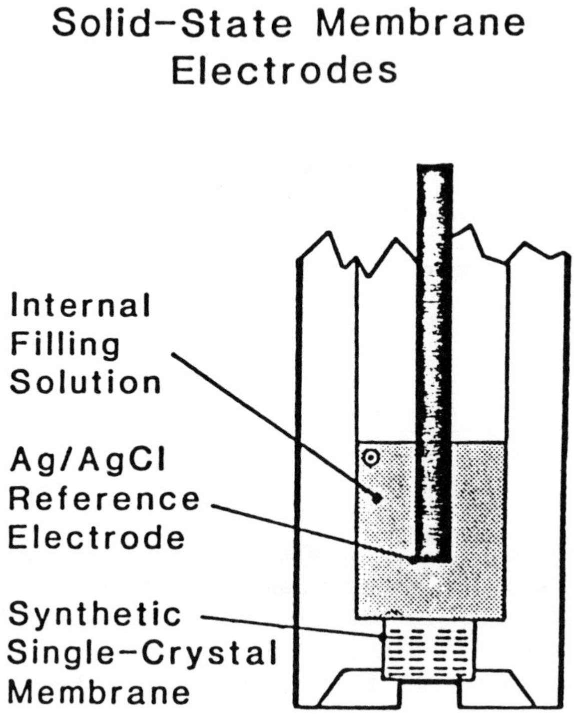

FIGURE 19

do wonders depending on what that liquid is or what you have tied to the walls of the porous membrane. A variety of things have been tied to the walls — enzymes, whole cells, and reactive species of all sorts, to take something in solution and convert it into another species that you can use to develop a useful signal. You can enzymatically convert proteins to make ammonia, or sugar to make carbon dioxide. Then you can use an ammonia detector, a CO_2 detector, or an oxygen detector. You can convert something to a specific, desired product, then into a useful signal. The list of "somethings" is quite long, as you can see in Figure 21. Most of the "somethings" are not yet commercial. The commercial ones started with the hydrogen ion 80 years ago and now encompass a long, long list of all the alkaline metals and alkaline earths, and also the chlorine, fluorine, bromine series. Several gas membranes are commercial — oxygen, CO_2, and some others. It is fairly easy to make membrane electrodes but it is hard to make them stable and long-lived. They work just fine on Tuesday, but on Thursday you are not getting the same signal

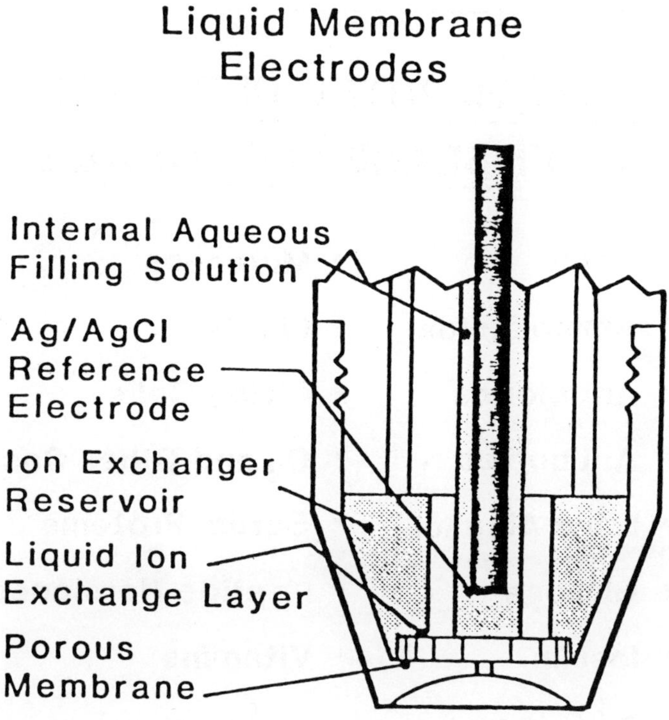

FIGURE 20

anymore. Also, one does not tend to get the Nernstian response — i.e., you should get a 59-millivolt change in potential for every factor of 10 change in external concentration. Frequently these membranes do not give you that. Those impediments have kept most of the membrane electrode applications from becoming commercially fruitful. Leaders of this field in the United States are Gary Rechnitz from the University of Delaware and his students. There is also a man named Suzuki from Tokyo who has done outstanding work with membrane electrodes. I think we will see the day when you can take a drop of blood, put it on a dish, place a microelectrode in it, and determine what you have and what you do not have in terms of disease. That has not been done yet, but I think it will be done in the next five or ten years. That is all I want to say about electrodes.

I want to turn now to liquid membranes. Liquid membranes are also not yet commercial. But I want to spend a few minutes talking about liquid membranes. I am not sure anyone said it explicitly in the last

SPECIES REPUTED DETECTABLE BY MEMBRANE ELECTRODES

- Acids
- Amino Acids
- Antibiotics
- Antibodies
- Ethyl Alcohol
- Glucose
- Insulin
- $>$ 20 Ions
- Mutagens
- Lipids
- Living Cells
- O_2 and Other Gases
- Serum Proteins
- Peptide Hormones
- Vitamins

FIGURE 21

couple of days, but it is certainly true that there are two key problems with synthetic polymeric — that is, dense — membranes: 1) they tend to be reasonably non-selective, in spite of what you may hear, in certain applications, and 2) they tend to be reasonably slow. They are non-selective because the things you want to separate are, in nature, frequently hard to separate, being chemically closely related; those separations are the toughest to make. Conventional polymeric membranes are not highly permselective. If they are permeable to A, they are permeable to B; and if A and B are chemically closely related — sorry. So synthetic polymeric membranes are very important, but, with a few exceptions, they are not very highly permselective and they are not very fast. By fast I mean the flux flow is quite low because they are dense. Diffusivities in solid polymeric materials are fairly small, on the order of 10^{-7} to 10^{-10} cm^2-

sec. For comparison, liquids tend to be 5 to 6 cm^2-sec. Gases tend to be 10 to 10^{-1} cm^2-sec. A projected solution to these flux and selection problems is to make liquid membranes. Figure 22 shows three kinds of membrane transport — and I think these are fairly universal — that one can conceive of. All membrane separation processes that I am aware of today operate by some transport mechanism. Take a solid or liquid membrane — all of the commercial ones are solid — and something passes from a region of high thermodynamic activity to a region of low thermodynamic activity. You hope that more of the stuff you want and less of the stuff you do not want passes through, and that a separation is thereby effected. This transport activity can be driven by a concentration gradient, by pressure differences, by electrical differences, or by even more exotic forces. There are two ways one can improve on that fairly simple approach: one way the world calls "facilitated transport," and the other way we call "coupled transport."

Let me take them in order. One can put inside a liquid membrane a specific carrier for the thing one wants to transport across the membrane. The example shown in Figure 22b is hemoglobin, which reacts specifically with oxygen. A man named Scholander, at Scripps Institute 24 years ago, carried out an experiment like this. He immobilized hemoglobin, which is fairly unstable outside the body, but in an experiment lasting only a few minutes he showed that he could increase oxygen flux by a factor of 10 using hemoglobin as the carrier. It works like this. Hemoglobin reacts with oxygen at one membrane/solution interface to form oxyhemoglobin, which diffuses from left to right across the membrane as shown here. The oxygen is stripped from the membrane at lower oxygen pressure, producing a reasonably pure oxygen stream. The hemoglobin then diffuses from right to left, so the feed side of the membrane picks up more oxygen. This process just keeps going like a shuttle. That is called "facilitated transport." I am not sure who the father of that term is, but it goes back a long way. Several people have done some good research in that area over the last generation, but nothing is yet commercial on a major scale. So, one can improve on passive membranes by using a carrier. One can improve on that one notch further by using a carrier in which the flow of the carried species (in the case of Figure 22c, the copper ion), is coupled to the flow of some other species, ideally one much cheaper than the species you are trying to separate. As is shown in Figure 22c, we could have a mixture of copper and other ions, say Cu^{2+} and Fe^{2+}. We could separate and concentrate the copper from such a solution as follows. You use a specific kind of carrier, in this case an oxime (HR). Since it is monovalent, it will react to form CuR_2 to the exclusion of other divalent metal ions. That complex diffuses from left to

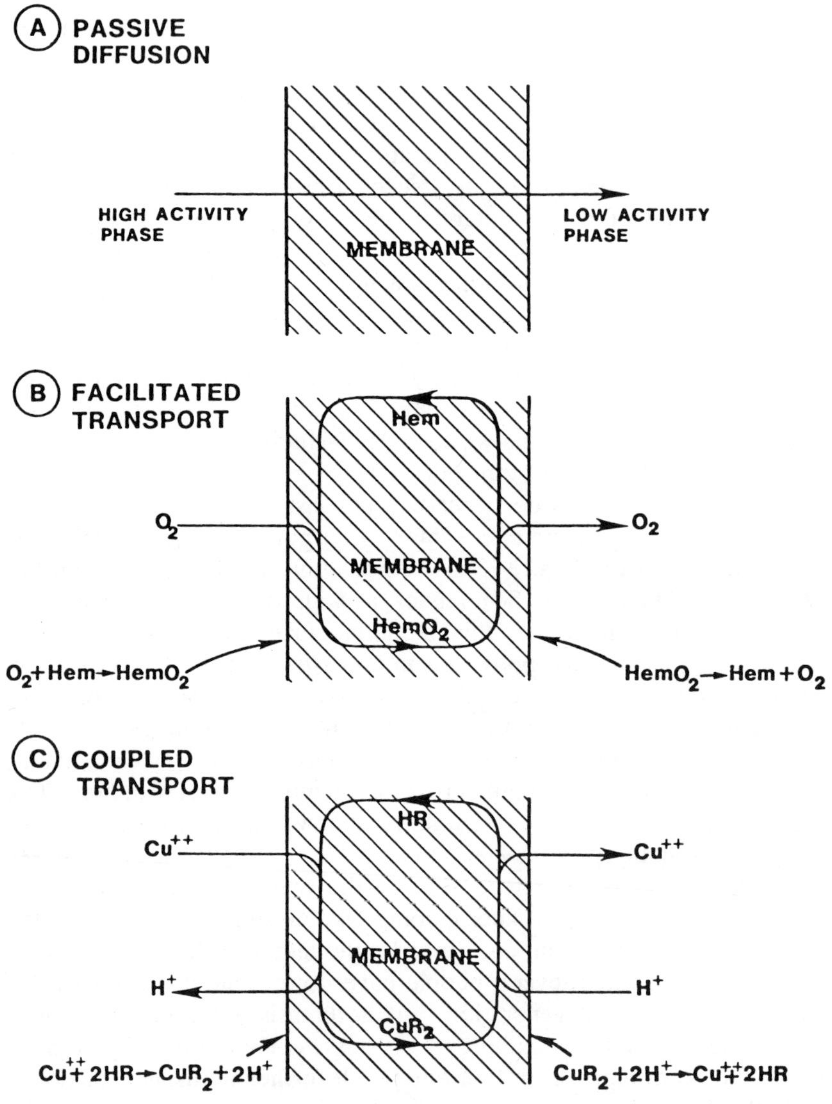
A PASSIVE DIFFUSION
HIGH ACTIVITY PHASE
MEMBRANE
LOW ACTIVITY PHASE
B FACILITATED TRANSPORT
Hem
O2
MEMBRANE
HemO2
O2
O2+Hem→HemO2
HemO2→Hem+O2
C COUPLED TRANSPORT
HR
Cu++
Cu++
MEMBRANE
H+
H+
CuR2
Cu+++2HR→CuR2+2H+
CuR2+2H+→Cu+++2HR

FIGURE 22

right across the membrane, shown here, dumps the copper at a lower pH, picks up hydrogen, and the HR diffuses from right to left. So again it is a shuttle, but it is a pump as well, because the only way copper can go from left to right to maintain electrical neutrality (and voltage is an extremely powerful driving force) is if hydrogen goes from right to left. You can turn that around because hydrogen is the driver. The only way hydrogen can go from right to left is to pump the copper from left to right. You can actually make copper go “uphill”. At least uphill externally; it is not going uphill within the membrane, because that would violate somebody’s law, probably the second law of thermodynamics, but it can certainly go uphill external to the membrane. It can go against its concentration gradient because hydrogen is going downhill, and because hydrogen has a much steeper concentration gradient, it can force copper to go uphill. We can use that approach to separate and concentrate a variety of things.

I want first to talk about facilitated transport and our attempts to use that to separate oxygen from air; then I am going to talk about metal-ion transport to separate and concentrate metals from solution. The advantages of liquid membranes with carriers are summarized in Figure 23. First, as I have said, you can demonstrate uphill diffusion; things will go against their concentration gradients by 10,000-fold or more if you do it right. It sounds impossible, but it works, and I have some data to prove it. Second, you can exhibit very high selectivity. These carriers are specific for the things you are trying to carry. You can separate things much more cleanly with liquid membranes than you can solid membranes. Third, they are fast. Now we are talking about diffusivities in the liquid phase rather than in a solid polymeric phase. Those are some of the advantages if you have the right kind of carrier. There are a couple of disadvantages, the most important of which is membrane life. We have concentrated our efforts at Bend Research over the past 5 to 6 years on improving membrane life, and with some success, I think.

LIQUID MEMBRANES WITH CARRIERS

1. "Uphill" Diffusion: $> 10^4$ X.

2. Extreme Selectivity: Separation Factors $> 10^3$.

3. Fast: $D \simeq 10^{-6} cm^2/sec$.

FIGURE 23

Liquid membranes are fairly simple to make. You take a microporous material such as the microporous polypropylene made by Celanese, under the trade name "Celgard," immerse it in a liquid — kerosene perhaps — and the organic liquid is drawn up by capillarity into the pores of the membrane, where it is reasonably stable. In fact, if the pores of the membranes are small enough, the membranes will withstand a differential pressure of a few atmospheres, which is enough for most applications. I know Bill Koros talked about the oxygen/nitrogen separation briefly. Figure 24 is a plot of the oxygen/nitrogen separation factor, basically the ratio of oxygen permeability to nitrogen permeability, of a series of membranes, as a function of the oxygen permeability. It is sort of A over B against A, if you will. There is a difference of 8 orders of magnitude in oxygen permeability from the glassy polymers to silicone rubber and to natural rubber. The thing I want to point out is that even though the permeability varies over a very, very wide range, you never get a very high selectivity. The highest selectivity reported is about 6. At a Gordon Conference last week I heard someone report a selectivity of oxygen over nitrogen of 100 — I don't believe it! Check the apparatus, would be my

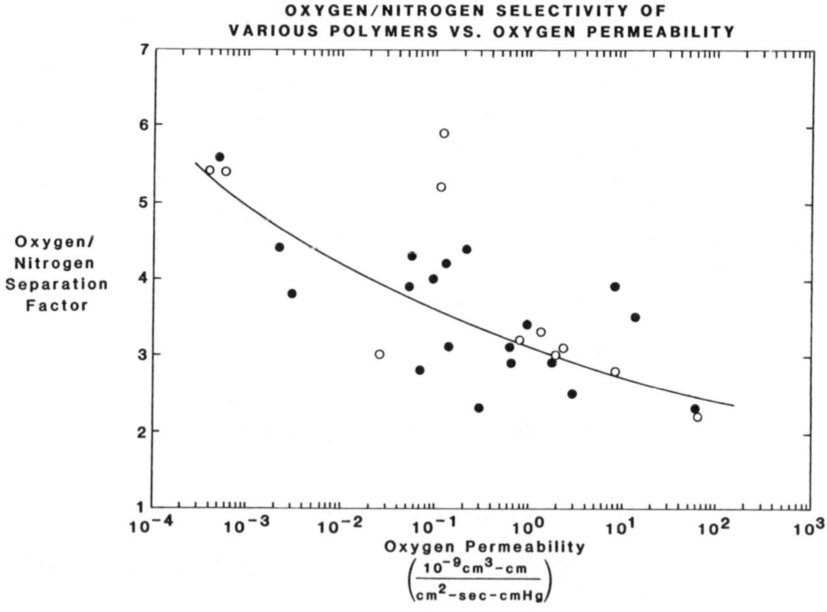

FIGURE 24

advice. Anyway, with that possible exception, which I won't accept right now, it is hard to separate oxygen from nitrogen because they are so chemically closely related and similar in size. Finding new polymers doesn't seem to be the solution, because if you did find something with a separation factor of 10 or 15, which might be the outer limit, the permeability of the oxygen is going to be extremely small, which means the flux of the membrane is going to be very low. Oxygen is a very cheap commodity; in the United States it sells for about $0.01 a pound, or about $20 a ton, so it is unlikely you can make a membrane thin enough and selective enough to make oxygen at the right price. We believe that polymeric membranes won't work for that separation. That has been proven around the world by ourselves and by most of the gas companies. But if you use a membrane with a carrier, as in our work, there is some hope of improving upon that. We have got selectivity that is as high as 50 in the laboratory so far, using a carrier — a hemoglobin analogue of low molecular weight that is stable for several weeks or several months. The carrier picks up oxygen, transports it down a pressure gradient — it's only downhill diffusion — and produces a reasonably enriched oxygen stream. I think the best we've done is something like 93%. Figure 25 shows nitrogen being rejected, but that is not quite true, because the solvent you dissolve the carrier in is not totally impermeable to nitrogen. Nitrogen

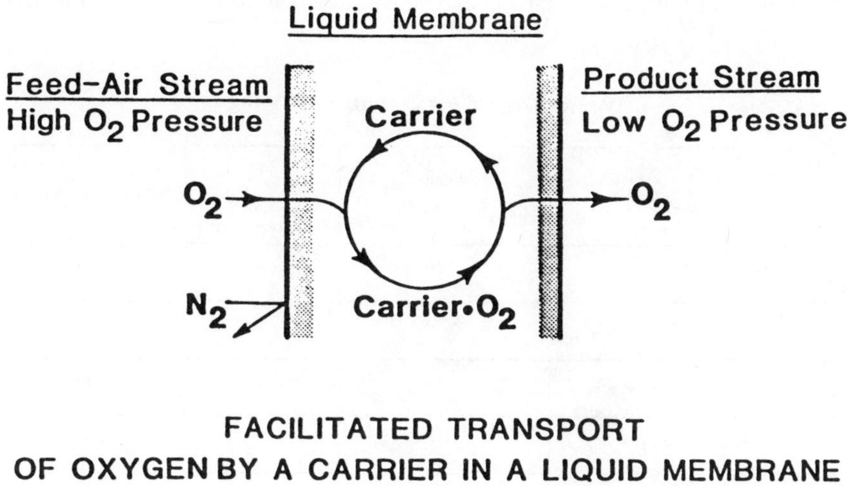

FIGURE 25

will diffuse by passive diffusion through this solvent, so you do not get a completely clean separation, but you rarely do with membranes.

Some data are shown in Figure 26 contrasting two very different kinds of polymers. Silicone rubber is the most permeable polymer known to man. It has an oxygen permeability of 60 in the units shown here. But the oxygen/nitrogen selectivity is only 2. Silicone rubber is only twice as permeable to oxygen as it is to nitrogen, so the highest-purity oxygen you can make is only about 36%, and that grade of oxygen doesn't have very many uses. Cellulose triacetate, on the other hand, has a very high selectivity, with numbers close to 6, and you can make a purer oxygen stream of 60%. But cellulose triacete exhibits a pretty low flux because the permeability is about 3 orders of magnitude lower than that of silicone rubber. You would need an extremely thin film — I think an impossibly thin film — to make oxygen. A couple of the membranes that we developed, in fact a whole family of them, have these properties. Permeability is comparable, in some cases, to silicone rubber; selectivity is about an order of magnitude higher than that of silicone rubber and 5 times higher than that of cellulose triacetate. These membranes have a lifetime of a few thousand hours. Not as long as one would like, but quite good.

I want to talk now about metal separation using liquid membranes with carriers. Figure 27 shows some data we obtained some time ago on the separation and concentration of copper across a liquid membrane that probably contained LIX 64. The X-axis, shown here as ΔpH, is the pH

OXYGEN/NITROGEN PERMEABILITIES

Membrane	Oxygen Permeability $\left(10^{-9} \frac{cm^3\text{-}cm}{cm^2\text{-}sec\text{-}cmHg}\right)$	O_2/N_2 Selectivity	Maximum Oxygen Concentration Produced (% O_2)
Silicone Rubber*	60	2.1	36
Cellulose Triacetate*	0.12	5.9	60
Bend Research Membrane No. 1	56	26	86
Bend Research Membrane No. 2	26	25	86

*S.-T. Hwang, et al., Separation Science 9 (6) 461 (1974).

FIGURE 26

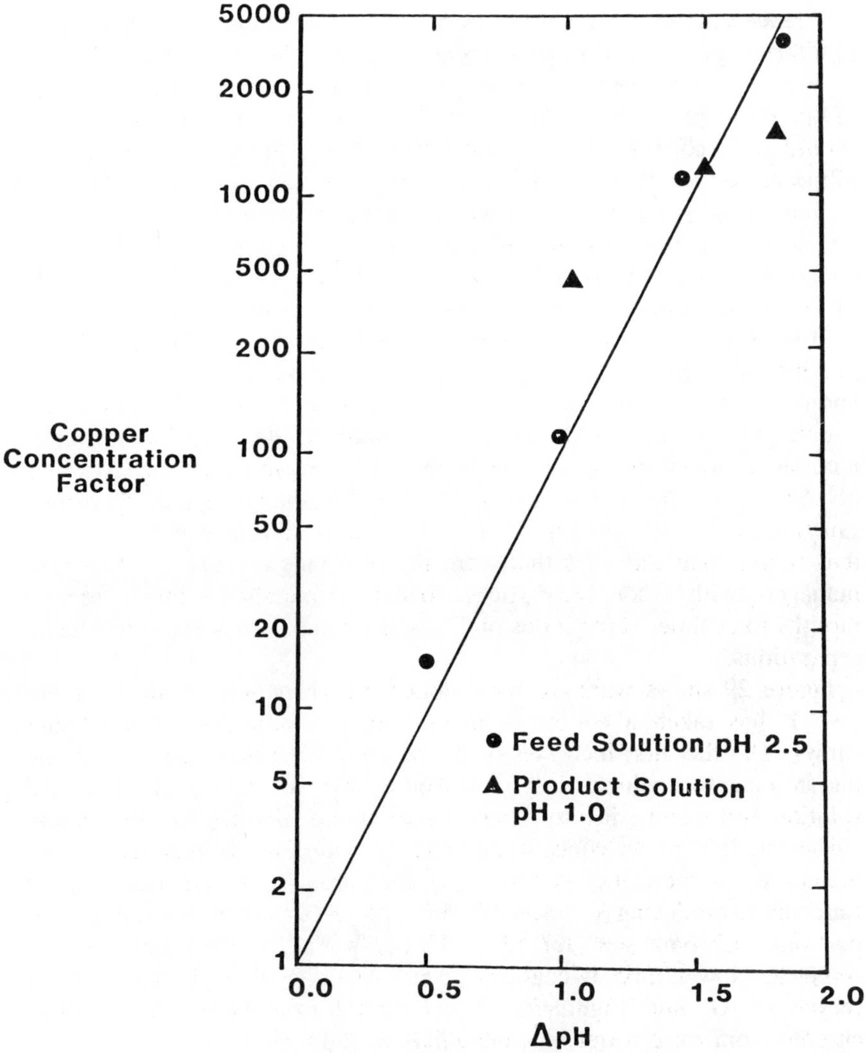

Steady state copper concentration factor vs. pH difference across the membrane. The line is from theory.

FIGURE 27

difference across the membrane. A pH difference of one unit is equivalent to a factor of 10 in hydrogen ion concentration. The Donnan Distribution Law, now 70 years old, says that with a divalent ion and a pH difference of one unit (that is, a 10-fold hydrogen ion concentration difference), you should get a copper concentration factor of 10^2, and you do. With a pH difference of 2 — that is, a 100-fold hydrogen ion concentration difference — you should get 100^2, or a 10,000-fold concentration of copper, and you do. So in fact, it does follow the Donnan Distribution Law and you can concentrate things many-fold with the right sort of pumping action. In this case we pumped it with hydrogen ions. What has been the problem with all metal separations using liquid membranes is membrane stability. We have concentrated on improving the stability, and Figure 28 shows about the best result we have obtained to date for copper-selective coupled-transport membranes. If you want to do a quick economic analysis, a pound of copper sells for about $1, so this is about $20 worth of copper per square foot per year. The flux shown in this figure is a function of time for, perhaps, 4 months. The membrane is not stable, but it is better than the ones that were in our hands a year ago. We have managed, with a couple of tricks, to extend membrane life to several months for copper separations and a year or more for some other metal separations.

Figure 29 shows what we have done with chromium in an 18-month test. It has taken a couple of membrane regenerations, as the figure shows. The flux first increases to a maximum then decreases — not the nicest thing to engineer a system around. We are taking chromic acid solution and moving it into a basic solution and winding up with sodium chromate. It is a very concentrated chrome solution. We regenerated the membrane periodically — about every 150 days. We regenerated it about four times, producing a reasonable flux of 5 to 10 pounds per square foot per year. Chrome sells for $2 to $3 per pound, so one can afford a somewhat lower flux. It could be useful in things like plating shops to recycle the chrome. Figure 30 is a photograph of a test we ran recycling chrome from rinse water in a metalplating shop.

Figure 31 shows the apparatus we used for uranium separation from vanadium. This was an early test. On the left-hand side of the cell was a solution at pH 1 and on the right was a solution at pH 12. In the middle was an immobilized-liquid membrane containing Alamine, our transporting agent. At 12:00 noon, we started the cell off with a solution containing 2000 ppm uranium and 100 ppm of vanadium on the left hand side. The separation of uranium and vanadium, if you recall your inorganic chemistry, is a pretty tough separation to effect. By about 2:00 we had separated most of the uranium from the vanadium. Almost all of the

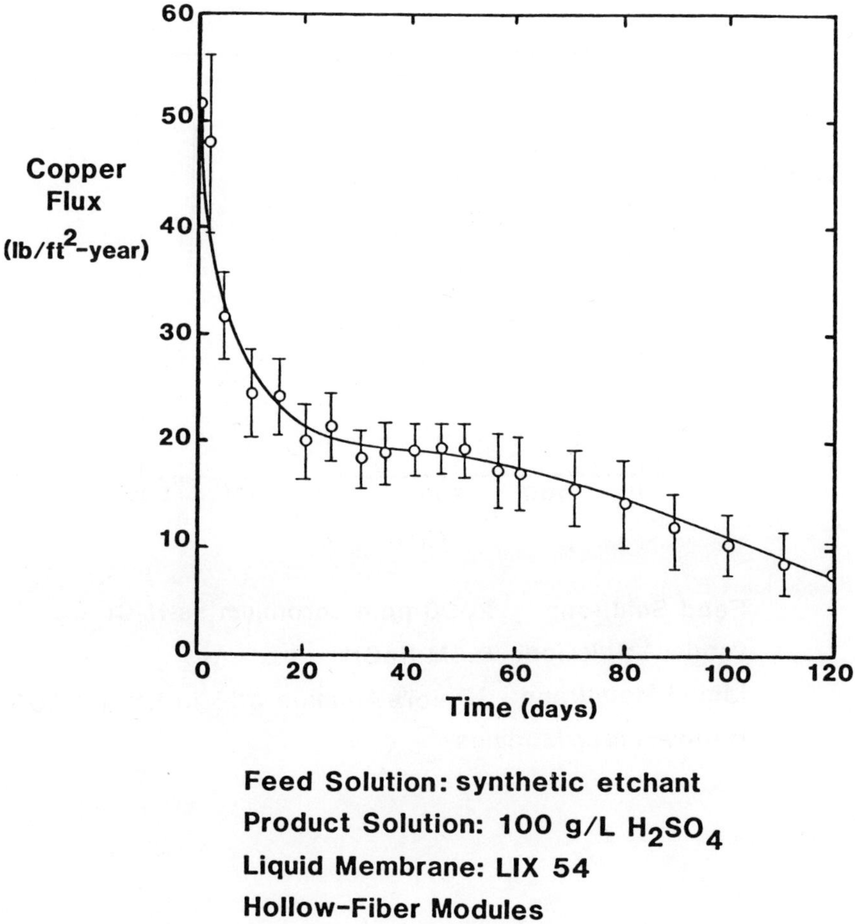

FIGURE 28

uranium was on the right and the vanadium was left behind. Vanadium was undetectable by our atomic absorption spectrophotometer in the product solution. And I would point out that after 1:00 the uranium was going uphill. There was one point when the uranium concentration was 1000 ppm on both sides, and after that the uranium was going uphill. Toward the end of the experiment uranium was concentrated on one side by 200-fold over the other side. One could go even longer and get even higher concentration factors, depending on the pH difference. So, one

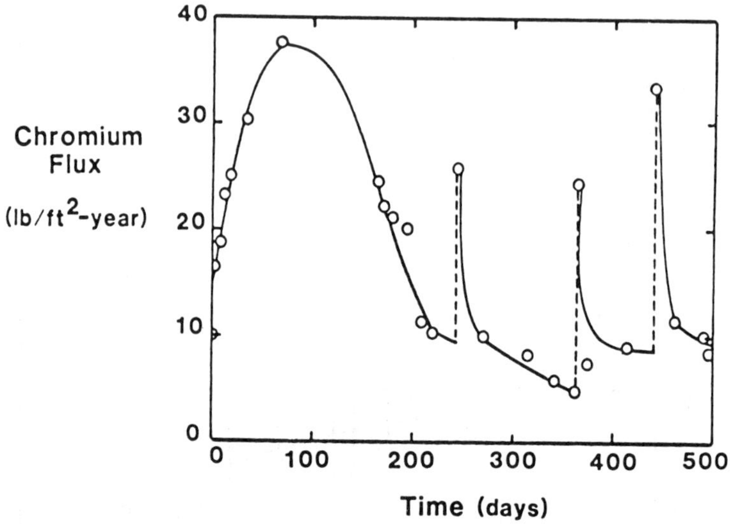

Feed Solution: 2000 ppm chromium as $H_2Cr_2O_7$
Product Solution: 0.1M NaOH
Liquid Membrane: 30 vol% Alamine 336 in Escaid 200
Hollow-Fiber Modules

FIGURE 29

can separate and concentrate uranium using the proper procedure. Figure 32 shows the results of a long-term test of the uranium-separating membrane. Flux was almost 20 pounds per square foot per year after nearly a one year without membrane regeneration. Figure 33 shows a small unit we built to test in Wyoming last summer that processes a few thousand gallons of water a day and extracts several pounds of uranium per day from a very dilute leach liquor. The extracted uranium is a precursor to making yellow cake. We can make about a 3% uranium solution from a 200-ppm feed of leach liquid with that small unit. There are four, 100-ft^2 hollow-fiber modules in the unit used for that separation.

FIGURE 30

It is not yet commercial, but we expect that it will be. The uranium market, as you may know, is a pretty sick market. So is the copper market. The whole mining industry is pretty poor around the world right now. But if it ever comes back we are waiting in the wings with some good technology. Figure 34 shows work done by Norman Li at Exxon that uses a different geometric approach to making liquid membranes. These are not immobilized, but rather mobile liquid membranes. We call this in our company the Norman Li approach and it has 20 or 30 patents associated with it. The process is to take little bubbles, as it were, made with surfactants, and dope that surfactant wall with a carrier, as we do with our membranes and as we do with immobilized-liquid membranes. Copper from a mixed divalent-ion solution would be concentrated in the sulfuric acid solution. The hydrogen ions would be pumped out and one can effect the concentration separation there. Ammonia from amines can be concentrated that way. Hydrogen sulfide and other organic acids will

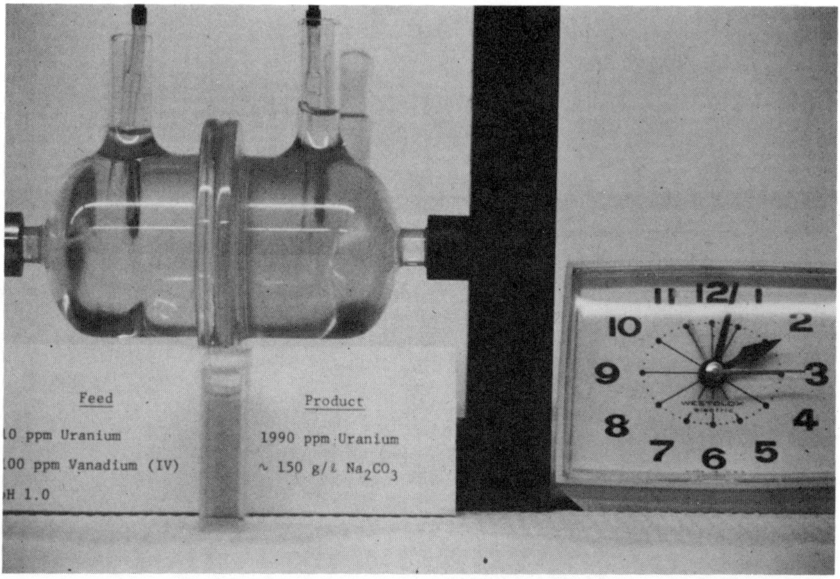

FIGURE 31

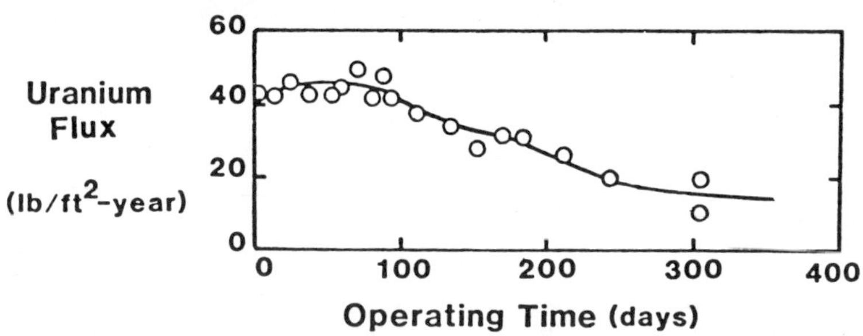

Feed Solution: 2000 ppm uranium as uranyl sulfate, 10 g/L H_2SO_4

Product Solution : 150 g/L Na_2CO_3

Liquid Membrane: 30 vol% tri-n-dodecylamine in Kermac 470B

Hollow-Fiber Modules

FIGURE 32

FIGURE 33

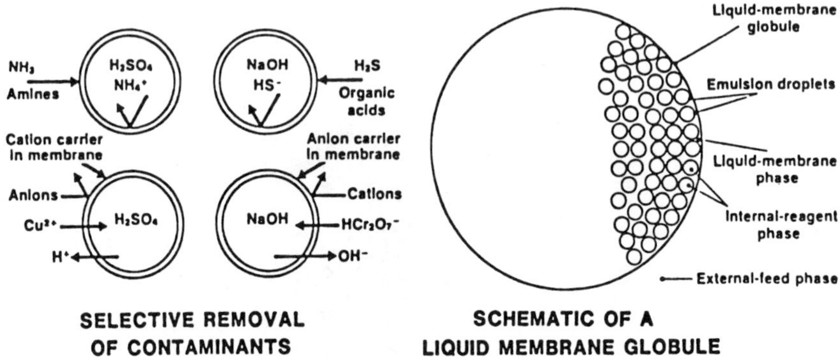

FIGURE 34

FIGURE 35

be in the basic solutions and so on. There are several applications, about two or three dozen by now, from that Exxon work. All these little bubbles are held in a liquid membrane globule just because that is the way nature is, and those separations are then carried out on those globules. Their problem, like ours, has been membrane stability. If any of these little bubbles break, you get dumping of what is inside to the outside world. And that is not good, obviously. So making stable globules and then unmaking them when you want to recover the product has been the problem with these mobile membranes. It seems like everyone has this problem. I do not think this process is commercial in the U.S. yet, but a great deal of work is being done on it. I think it is also technology waiting in the wings for the right sort of price in the metals market, which could come if the EPA gets tougher about enforcing pollution control.

Figure 35 shows our research lab, and there is a small manufacturing operation downtown as well that makes gas separation modules. That is the end of my talk. Thanks for your attention.

DISCUSSION

QUESTION Can you elaborate on any of the tricks to extend the lifetime of the liquid membranes?

ANSWER Tricks I am not going to tell you, and I did not hear the first part of your questions.

QUESTION What was the liquid in the oxygen-nitrogen membrane?

ANSWER It was a cobalt complex. Actually, it is now patented so I can tell you something about it. It is dissolved in the right sort of solvent. In fact the trick to making it work is not so much the carrier you use, but the environment you put it in. It is confusing; if you really want to know more I can send you the patent. There are several things in there. They key element is a Cobalt Schiff base of MW 300 or 400, something like that. It has a reasonably long lifetime. The trick is getting something that will partially oxidize. Cobalt exists in two oxidation states. You do not want it to get permanently oxidized to cobalt(III), which is not an oxygen carrier. There is some sort of permanent, irreversible oxidation of cobalt(II) to cobalt(III). The same thing occurs with the other transition metal carriers such as manganese, iron, and so on. We found one that, under the right conditions, has a pretty long life. Tricks! The family jewels! No way will I tell!!

QUESTION What is it that limits the life of the liquid-filled membranes? Evaporation of the liquid, or loss of the liquid, or is it breakdown of the carrier?

ANSWER It is loss of the liquid, but not by evaporation. Once lost, the liquid gets displaced by the aqueous phase. It took us a long time to figure that out. We knew what was happening, but didn't know why it was happening. If you know why it happens you can figure out the solution, obviously. There are several approaches that we have explored. It is displacement of the liquid. That is probably why the flux goes up. The membrane is getting thinner.

Membrane Research and Applications in Japan

SHOJI KIMURA

Institute of Industrial Science, University of Tokyo, 7-22-1 Roppongi, Minatoku, Tokyo 106, Japan

Japan Society of Membrane Science was established in 1977 and it has been publishing a bimonthly journal "Maku(membrane)". One of it's objects is to unite people in both artificial and biological membranes for mutual understanding, which may lead to the development of new membranes and technologies. Now about half of it's members are biological and medical people. Many other societies, such as the Society of Polymer Science, the Society of Chemical Engineers and Japan Chemical Society are holding many meetings and symposiums on membranes every year.

The Ministry of International Trade and Industry (MITI) is now developing a program called "Research and Development Project of Basic Technology for Future Industries", in which R & D of Synthetic Membranes for New Separation Technology is included. This project was started in 1980 and will last until 1990. MITI has another project called "Large Scale Project", in which 8 projects are included and one of which is a so-called "C. 1 Chemistry." Here R & D of catalyst and gas separating membranes are underway.

The long term running test of a reverse osmosis desalination plant, whose capacity is 800 m^3/day, has been performed since 1979 by the Water Reuse Promotion Center (WRPC) under the MITI. WRPC has also been operating many other desalination and water reuse projects using reverse osmosis and ultrafiltration technique.

Projects of other ministries and fields of applications will be presented at the symposium.

Membrane research activities in Japan are now wide spread over many fields and at least four ministries are now concerned with them. In the following the present status will be explained mainly based on five papers presented at Europe–Japan Joint Congress on Membranes and Membrane Processes held on June 18–22, 1984 at Stresa, Italy[1,2,3,4,5].

I. ACTIVITIES OF VARIOUS SCIENTIFIC AND ENGINEERING SOCIETIES

Japan Society of Membrane Science was established in 1977 and it has been publishing a bimonthly journal "Maku(membrane)". Other activities are listed in Table I. One of its objects is to unite people in both artificial

TABLE I

Activities of Japan Society of Membrane Science (Established in 1978, President: Prof. M. Nakagaki, Kyoto University)

1 *Publication of the bimonthly journal "Maku (membrane)"*
volume 9 in 1984

2 *Annual meeting in May every year*
about 60 papers on artificial and biological membranes and symposium or plenary lecture
recent subject of symposium:
Artificial cell membranes
Renal membranes and its simulation

3 *Lectures on special topics in every 3 month*
Recent lectures:
Present status of pervaporation process
T. Ishikawa (Tokyo Metropolitan Univ.)
Gas separation by inorganic porous solid membrane
S. Asaeda (Hiroshima Univ.)

4 *Courses on membranes and membrane processes in Nov.*
Coming course is "Process design of various membrane separation processes"

5 *Lectures of foreign guests at appropriate time*
Recent lecture: A general overview of Dow Chemical's membrane development by Dr. E. H. Wagener

6 *Europe–Japan Joint Congress on Membrane Technology*
at Stresa, Italy in June, 1984. Next in 1987 in Japan

and biological membranes for mutual understanding, which may lead to the development of new membranes and technologies. Now about half of its members are biological and medical people. Many other societies, such as the Society of Polymer Science, the Society of Chemical Engineers and Japan Chemical Society are holding many annual meetings and symposiums related to membranes every year.

II. R & D ACTIVITIES OF VARIOUS MINISTRIES

Activities of various ministries are listed in Table II. The Ministry of International Trade and Industry (MITI) is spending the largest amount of budget on four projects. Many national universities are under the Ministry of Education, whose budget is not so large comparing numbers of researchers. Usually cooperation between the different ministries are not good.

TABLE II

National R & D Activities for Scientific & Industrial Membranes (1983)

Ministry	R & D Project	Member (people) National Labs.	Private Industry	University	Budget (10^8 Yen) (10^6\$) Full Support	Partiall Support
MITI	Basic Technology for Future Industries	3(28)	9(41)	1(1)	6.2(2.7)	—
	Large Scale Industrial Technology (Cl Chemistry)	1(5)	3(10)	0	5.1(2.2)	—
	Oxygen Enrichment	0	7(20)	0	—	4.0(1.7)
	Water Re-use Promotion Center	—	5(14)	—	—	1.2(0.5)
MAFF	Membrane Technology for Food Industry					1.1(0.4)
STA	Exploratory Research for Advanced Technol.				1.3(0.5)	
ME	Ordinary R & D			?(298)	10 (4.1)	
	Sum	4(33)	24(85)	?(299)	12.6(5.4)	6.3(2.6)

MITI: Ministry of International Trade & Industry,
MAFF: Ministry of Agriculture, Forestry & Fisheries,
STA: Science & Technology Agency,
ME: Ministry of Education

III. MITI'S PROJECTS

The largest project is the basic technology for future industries, which consists of three main themes as listed in Table III, where the membrane project called "Synthetic Membranes for New Separation Technology" is included. Organizations participating in this project are listed in Table IV, its 10 year program is in Table V, and its main subjects are in Table VI. In these tables meaning of some words's somewhat obscure, because the project is now under development and it is premature to disclose interim results.

The next project is one of the Large Scale Industrial Technologies, and called "C1 chemistry". The object of this program is to use CO and H_2 as starting materials for chemical industries in Japan, because we have no petroleum and natural gas resources. Examples of schemes of C1 chemistry are shown in Table VII. The schedule and budget are shown in Table VIII. Main budget has been used for catalyst development and the

TABLE III

Research and Development Projects of Basic Technology for Future Industries

1 *New Materials*
High performance Ceramics
Synthetic Membranes for new Separation Technology
Synthetic Metals
High Performance Plastics
Advanced Alloys with Controlled Crystalline Structure
Advanced Composite Materials
2 *Biotechnology*
Bioreactors
Large Scale Cell Cultivation
Utilizing Recombinant DNA
3 *New Electronics Devices*
Superlattice Devices
Three Dimensional ICs
Fortified ICs for Extreme Conditions

TABLE IV

Members of R & D of Synthetic Membranes for New Separation Technology

National Laboratories;
National Chemical Laboratory For Industry
Research Institute For Polymers and Textiles
Industrial Products Research Institute
Private Companies;
Asahi Chemica Ind. Co. Ltd
Asahi Glass Co. Ltd
Kuraray Co. Ltd
Sumitomo Electric Ind. Ltd
Daicel Chemical Ind. Ltd
Teijin Ltd
Toyobo Co. Ltd
Toray Ind. Ltd
Mitsubishi Chemical Ind. Ltd

rest for the development of membranes separating gas mixtures, whose target values are in Table IX. Membranes under development are listed in Table X, some of which (for example polyimid membrane developed by Ube Ind.) are claimed to reach the target.

Membranes for enriching oxygen from air are being developed by a union of seven companies under the MITI, but details have not been disclosed.

As for the Water Reuse Promotion Center (WRPC), this is a nonprofit organization under the MITI and its aim is to develop technologies for

TABLE V

Program for Ten Year term

PHASE I (1981–1984)
Clarification of Separation of Membranes
Primary and Higher Structure
Polymer Functionality for Permselectivity
Screening of Polymers

PHASE II (1985–1987)
Development of Membrane Materials and Structure
Development of New Design of Modules and Systems

PHASE III (1988–1990)
Fabrication of High Performance Membranes and Systems
Scale-up Evaluation

TABLE VI

R & D Subjects for Membranes

1 *Common Basis for Membrane*
Separation Processes, Molecular Designing for Membrane, New driving Force, etc.

2 *Membrane Synthesis*
Affinity Gradient Membrane, Plasma Polymerization Membrane, Bio-Mimetic Membrane, Hollow Fiber, Asymmetric Membrane, Conformation Controlled Membrane, Surface Activity Controlled Membrane, Ultra-thin Layer Membrane, etc.

3 *Separation Mechanism*
Interaction Between Membrane Material and Permeant, Relationship Between Polymeric Conformation and Permselectivity, Addition of Advanced Function to the Membrane, etc.

4 *Membrane Property Evaluation*
Membrane for Liquid and Membrane for Gas

TABLE VII

Schemes of so-called Cl Chemistry

1 *Ethylene Glycol*
$2CO + 3H_2 = CH_2OHCH_2OH$ (1)

2 *Ethanol*
$2CO + 4H_2 = C_2H_5OH + H_2O$ (2)
$CH_3OH + CO + 2H_2 = C_2H_5OH + H_2O$ (3)

3 *Acetic Acid*
$2CO + 2H_2 = CH_3COOH$ (4)

4 *Hydrocarbon*
$2CO + 4H_2 = C_2H_4 + 2H_2O$ (5)
$2CH_3OH = C_2H_4 + 2H_2O$ (6)

TABLE VIII

Long-term Schedule and Budget of Cl Chemistry

	1980	1981	1982	1983	1984	1985	1986	1987
Gas Separation and Refining Technologies (Membrane Method)	Elementary Studies		Research and Development		Experiment with Bench Plant		Supports of Technologies	
					Design and Construction		Operation of Pilot Plant	
Budget (10^8 Yen)	0.05	0.57	3.72	5.07	3.83			
(10^6 US$)	0.02	0.24	1.55	2.11	1.60			
New Synthesis Processes of Basic Chemicals	Elementary Studies		Research and Development		Experiment with Bench Plant		Support of technoogies	
					Design and Construction		Operation of Pilot Plant	
Survey of Gas Generation Process and Design of Total Synthesizing Plant					Survey		Design of Total Plant	
Total Budget (10^8 Yen)	0.49	9.02	25.27	25.27	17.34			
(10^6 US$)	0.20	3.76	10.5	10.5	7.23			

R & D Period: 8 years, Total Budget: 15 billion Yen (62.5 million US$)

TABLE IX

Target Values of Gas Separating Membranes

	Pressure (kg/cm^2)	Max. Temp. (C)	Separation Factor H_2/CO	CH_4/CO	Permeability (cm^3 cm/cm^2 cmHg sec)
Porous Membrane					
organic	>50	150	2–3	—	$>10^{-4}$
inorganic		400			
Non-porous Membrane, organic	>50	100	>20	2–3	$>10^{-9}$

TABLE X

Materials and Types of Membranes Being Developed

Ube Industries, Ltd.	Nonporous, Asymmetric Polyimid Membrane Porous, Polyimid Membrane
Toyobo Co., Ltd.	Nonporous, Asymmetric Polysulfone Membrane Porous, Silica Glass Membrane
Sumitomo Electric Industries, Ltd.	Nonporous, Plasma Polymerized Membrane on Porous PTFE Substrate

water desalination and water reuse. For these purposes many membrane technologies have been developed. These are listed in Table XI. 800 m^3/day seawater desalination unit is still operated to perform the long-term test of modules. Its special feature is an energy recovering turbine combined with a motor and a pump and it recovers about 40% of energy from the rejected brine. Membranes used are spiral wound type PEC-1000 made by Toray Ind. Inc. and hollow fiber type cellulose acetate made by Toyobo Co., and they are tested alternately for one month.

A recent project is a production of boiler feed water from seawater by RO using a two stage desalination scheme using the above two types of module. Regarding the remaining projects listed in Table XI, results are reported in the annual reports from WRPC.

IV. MAFF'S PROJECT

A research development association of membrane technology in the food industry has been organized by a partial support from MAFF. 18 company pairs, each of which consists of a food manufacturer and a membrane manufacturer or an engineering company, are cooperating to

TABLE XI

Major Project for Membrane Process at Water Re-Use Promotion Center

Projects	'74	'75	'76	'77	'78	'79	Year '80	'81	'82	'83	'84	Note
Experiments												
1. Seawater Desalination by RO process	Imported Memb.			Domestic Memb.			800m^3/day Demo. Plant					Continued
							High Temp. & High Conc. Feed					
2. Re-Use of Drainage for Office Buildings												UF, RO
3. Advanced Water Treatment of Secondary Effluent for the Sewage Treatment Plants by Membrane Process							200m^3/day Pilot Plant					
4. Operational Data of a Small RO Seawater Desalination Plant in Takegashima Island												since '81 in service
5. Re-Use of Waste Water from the Washing Process of Spent Bone Charcoal in a Sugar Refinery												RO
6. Production of Boiler Feed Water from Seawater by RO Process												RO
7. Re-Use of Washing Waste Water in a Radiator Manufacturing Factory												RO
8. Re-Use of Washing Waste Water in an Aluminum Rolling Mill												RO
9. Re-Use of Waste Water in a Dyeing Factory of Cotton and Polyester Yarn												Continued
10. Development of a New Technology for Waste Water Treatment Using Bioreactor and/with Membrane System												Continued UF
Feasibility Studies												
1. Feasibility Study on the Construction of Seawater Desalination Plants in the RYUKYU Islands												Continued RO, MSF
2. Feasibility Study on the Construction and Practical Use of Seawater Desalination Ships												RO, MSF

perform research subjects listed in Table XII, which are divided into four categories. This association will be terminated in fiscal 1984 and reformed to a new organization.

V. MPHW'S PROJECT

As is shown in Table XIII the Ministry of Public Health and Welfare has been encouraging medical and pharmaceutical companies to equip water purification units using membranes due to their energy saving nature, and it has organized a union to authorize membrane equipment that can be purchased tax free. Another project has also been conducted by this union to verify the safety of membrane treated water for injection water. This result will lead to revision of the Japan Pharmacopoeia in the near future.

TABLE XII

R & D of Membrane Technology in Food Industry

Processing	Theme
Milk	1) Concentration of whole milk and skim milk by UF
	2) Concentration of cheese whey by RO
	3) Fractionation of whole milk or cheese whey by UF
	4) UF separation of inhibitors of microorganism growth for fermentation
	5) Concentration of fermented milk
	6) Concentration of milk by ceramic membrane
Fruits juice	7) Fractionation and clarification of citrus juice by UF
	8) UF and RO recovery of pectin in waste water from canned mandarin orange juice
	9) UF clarification and RO concentration of apple juice
Animal protein	10) Concentration of egg white by UF
	11) RO and UF recovery of protein in waste water from ham processing
	12) UF and MF separation of colloidali material and brown color material from molasses in bread yeast factory
	13) UF separation of emulsifiers and whipping agents from enzymatic treated soybean extract
Soybean protein	14) UF and RO separation of color materials and salts from hydrolysed vegetable protein (HVP)
	15) Concentration of soybean milk by UF and RO
	16) Removal of unpleasant smell components from soybean milk
	17) Clarification of raw soy sauce by UF
	18) Clarification of heat treated soy sauce by UF

TABLE XIII

Projects of Ministry of Public Health and Welfare

1) *Organization of Association for the Advancement of Membrane Technology (AAMT)* Authorization of the selection of membrane apparatuses, which can be purchased tax-free, recognizing their energy saving nature.
2) *Experimental test for the revision of Japan Pharmacopoeia about the water for injection* Measurement of toxic components in membrane and module materials Dissolution test of every materials used following FDA standards Measurement of rejections of bacteria and endotoxin using actually working units

VI. ME'S PROJECT

Numbers of university people engaging in membrane research and funds are listed in Table XIV, where it is seen about 100 people are working on artificial membranes.

VII. MEMBRANE MARKET AND APPLICATIONS IN JAPAN

The membrane market in Japan is given in Table XV. The largest one is dialysis membranes for artificial kidneys and the next is miscellaneous uses of microfilters.

One of the growing markets is production of ultra-pure water for the electronics industry and an example of the line is shown in Fig. 1, where RO is used for pre-desalination before an ion exchange tower and UF is used for removing fine particles before the use point.

The next interesting application is the concentration of lignin from BKP alkali effluent of pulping factories. Its flow scheme and various data is shown in Fig. 2 and Table XVI, respectively. The concentrated lignin is burned as a fuel and the permeate can be discarded without treatment, so that fuel economy and decrease of amount of sludge can be realized at the same time.

In large cities like Tokyo or Osaka big buildings are asked to equip water reuse systems to save the water supply shortage and reduce the amount of effluent. One example, that uses UF is shown in Fig. 3. Another example shown in Fig. 4 uses the combination of activated sludge reactor and UF. Sometimes it is safer to adopt the latter system to avoid the direct contact of effluent and membranes. It is also interesting that the amount of sludge to take out reduces very much in this scheme.

TABLE XIV

Research Fund of Japanese Universities Supported by the Ministry of Education on Membranes (1983), 10^6 Yen (No. Person)

A. Artificial Membranes	438.40(102)
A — 1 Transport Phenomena	73.25 (31)
Polymeric Membranes	61.30 (22)
Gas	6.50 (4)
Ions, Solutes	38.50 (14)
Sensor	7.9 (2)
Battery	2.0 (1)
Others	6.4 (1)
Liquid Membranes	11.95 (9)
A — 2 Others	365.15 (71)
Polymeric Membranes	39.40 (19)
Amorphous Inorganic & Metal Membranes	245.85 (28)
Liquid Membranes	35.2 (10)
Others	44.7 (14)
B. Biomembranes	560.02(197)
Transport Phenomena	79.60 (18)
Receptor	26.5 (10)
Energy Transfer	5.60 (2)
Others	448.32(167)
C. Total	998.42(299)

TABLE XV

Membrane Market in Japan (1983), Unit: 10^8 Yen (10^6 US$)

	Dialysis	Electro-dialysis	Electrolysis	R.O.	U.F.	M.F.	Others	Sum
Artificial Kidney	500							500(217)
Salt Production		10						10 (4)
Desalination				5				5 (2)
Chlor-alkali			?					?
Food, Fermentation, Pharmaceuticals				?		30		30 (13)
Water Treatment					5			5 (2)
Pure Water for Electronics				10	5	30		45 (20)
Battery and others				?		40	?	40 (17)
Sum	500	10	?	15	15	100	?	640(277)

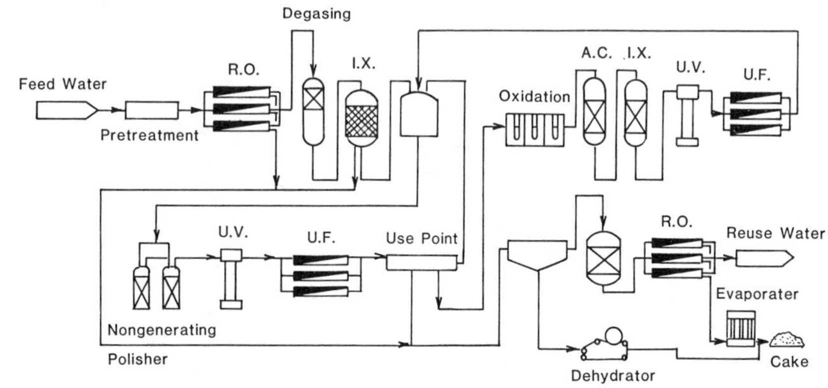

FIGURE 1 Ultra-pure Water Production Line.

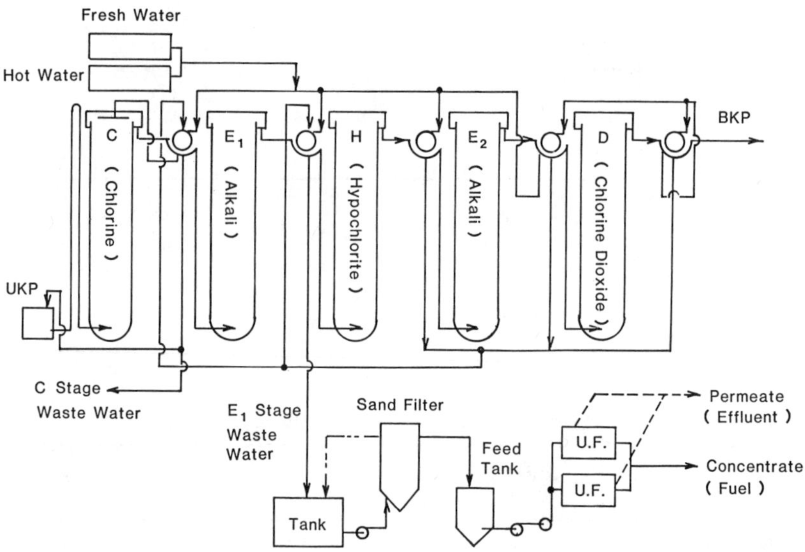

FIGURE 2 Recovery of BKP Wash water by U.F.

TABLE XVI

Data of BKP Alkali Effluent Treatment by UF (50°C)

	Designed Values	Actual Values	Permeate	Concentrate
Capacity (m^3/day)	4,000	3,600	3,382	218
Concentration Ratio	20	16.5		
COD (ppm)	1,250	1,900	430	24,700
pH	10.5	10.5	10.5	10.5
Color (ppm)	—	6,500	440	130,000
TS (ppm)	—	6,700	4,600	39,000
SS (ppm)	100	100	0	1,650

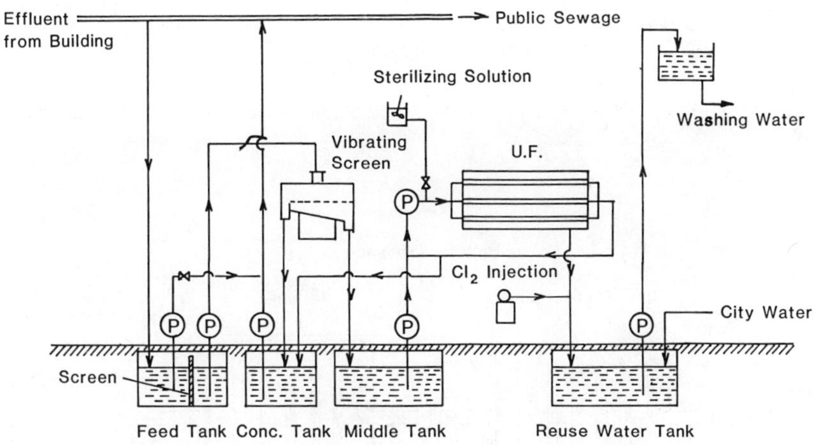

FIGURE 3 Water Reuse Process Using U.F. (Mitsubishi Rayon Eng. Co.)

MITI is now planning to improve and expand this scheme to larger effluent treatment plant. This plan is called "Aqua-Renaissance '90" and as shown in Fig. 5, it tries to combine bioreactors and membranes.

VIII. MEMBRANE MANUFACTURERS IN JAPAN

Membrane manufacturers and their products are listed in Table XVII.

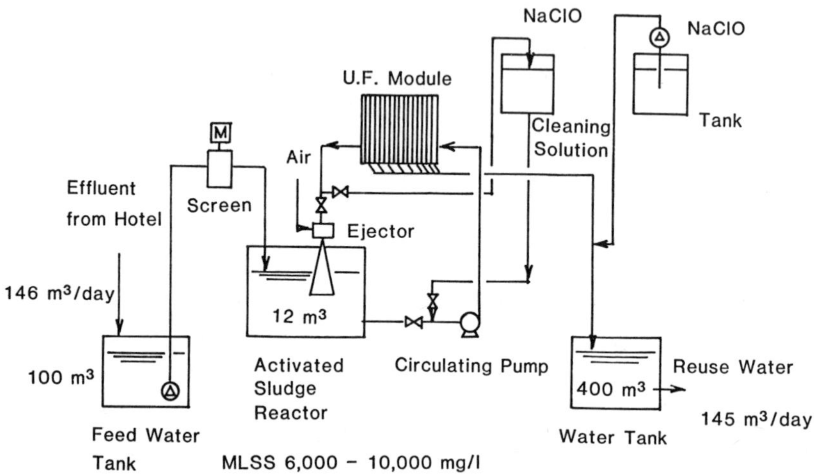

FIGURE 4 Water Reuse System with Activated Sludge Reactor and U.F. (Miyako Hotel, Tokyo).

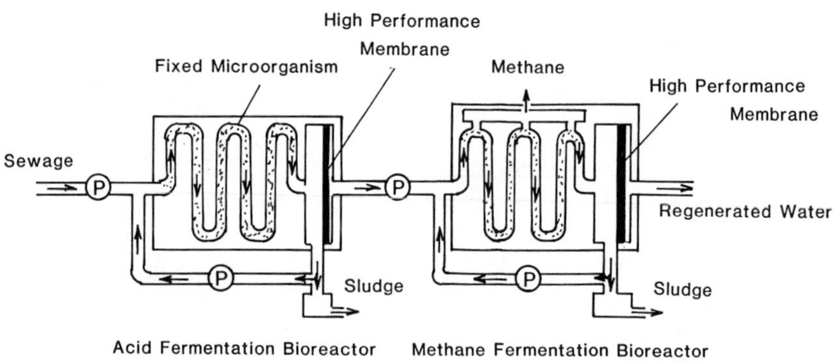

FIGURE 5 Aquarenaissance '90.

TABLE XVII-1

List of Membrane Manufacturers in Japan (1984)

Manufacturer	Trade Name	Shape	Fractional/ Rejec-Mole. Wt./ tion (%)	Material	Flux $(1/m^2hr)(kg/cm^2)$
Asahi Chem. Ind. Co. Ltd	*UF*				
	ACV-3010	Capillary	13,000	PAN	177(1)(25°C, PW)
	-3050		13,000		135
	AIV-3010		6,000		36
	FCV-3010		13,000		219
	-3510Z		13,000		225
	SIP-3013		6,000	PS	212
	KCV-3013		13,000	PAN	26(1)(Cationic E.P.)
	MF				
	SR-205		0.1 m (Pore Dia.)	PO	1,500(1)(25°C, PW)
Daicel Chem. Ind. Ltd	*RO*				
	DRS-97	Spiral & Tubular	96–98(0.35% NaCl)	CA	3(40)(25°C, 0.5% NaCl)
	-95		94–96%		50
	-90		86–94%		75
	-70		60–80%		125
	-50		40–60%		63
	- 0		<5%		104
	UF				
	DUY-HH		5,000	PAN	60(3)(25°C, PW)
	-H		10,000		125
	-M		20,000		200
	-L		40,000		300
	DUS-40	Tubular	40,000	PES	200
Sumitomo Heavy Ind. Ltd	*RO*				
	CA-98	Tubular (Outside)	98%(0.5% NaCl)	CA	17(35)(25°C, 0.5% NaCl)
	-95		95%		37
	-90		90%		54
	UF				

TABLE XVII-1 (*Continued*)

List of Membrane Manufacturers in Japan (1984)

Manufacturer	Trade Name	Shape	Fractional/ Rejec-Mole. Wt./ tion (%)	Material	Flux ($1/m^2hr$)(kg/cm^2)
	CA-UF		3,000		29−37(4)(25°C, PW)
	PS-UF		16,000	PS	167−291
Sumitomo Chem. Co.	*RO*				
	SC-0200	Tubular	97.5−98.5%(0.5% NaCl)	PAN	60(50)(25°C, 0.5% NaCl)
	-0500		94−96%		75
	-1000		88−92%		90
	-2000		75−85%		120
Teijin Ltd.	*RO*				
	TL-198	Tubular	97−99%(0.5% NaCl)	PBIL	>50(42)(25°C, 0.5% NaCl)
	-195		93−97%		>65
	-290		75−95%(10% Sucrose)		>30(20)(25°C, 10% Sucrose)
	-250		40−60%		>40
	-230		20−40%		>50
	-215		10−20%		>60
Sumitomo Bakelite Co. Ltd.	*UF*				
	FPM-1000	Capillary	10,000	PS	200(1)(25°C, PW)
	-3000		30,000		500
	-8000		80,000		700
Nitto Electric Ind. Co. Ltd.	*RO*				
	NTR-1597-P18A	Tubular ($1.6m^2$)	97%(0.5% NaCl)	CA	28(42)(25°C, 0.5% NaCl)
	-1595-P18A		95%		33
	-1590-P18A		90%		43
	-1570-P18A		70%		38(20)
	-1550-P18A		50%		56
	-1530-P18A		30%		76
	-1510-P18A		10%		89
	-7199S4	Spiral	99%(0.5% NaCl)	PA	4 m^3/day(42)

	-7197S4	(7.6m^2)	97%(0.2% NaCl)		6.4	(30)
	-7250S4		50%	PVA	20	(20)
Nitto Electric	*UF*					
Ind. Ltd.	NTU-2020-P18A	Tubular	20,000	PO	53(2)(25°C, 0.5% PEG #20,000)	
	-20100-P18A	(1.6m^2)	100,000		142	
	-3020-P18A		20,000	PS	56	
	-3508-P18LP	(2.3m^2)	8,000		64(2)(25°C, 0.5% PEG #6,000)	
	-3520-P18LP		20,000		65(2)(25°C, 0.5% PEG #20,000)	
	-35100-P18LP		100,000		176	
	-4208-M4	(0.39m^2)	8,000	PI	64(2)(25°C, 0.5% PEG #6,000)	
	-4220-P18B	(0.75m^2)	20,000		60(2)(25°C, PEG #20,000)	
	-8050-C15	Capillary (5m^2)	50,000	PA	440(1)(25°C, PW)(m^3/day)	
	-3050-C3	(6.3m^2)	20,000	PS	270(1)(25°C, PW)(m^3/day)	
	-3150S4	Spiral (7.6m^2)	50,000		40(2.5)(25°C, PW)	
Kuraray	*MF*					
Co. Ltd.	SF-101	Capillary	0.02 m (Pore Dia.)	PVA	100(1)(25°C, PW)	
	-301		0.04		450	
	-501		0.1		400	
	UF					
	CFW-6021		20,000	PS	225	
	-4601		260,000		450	
Toray	*RO*				(m^3/day)	
Ind. Inc.	SC-1100	Spiral	93–95%(0.15% NaCl)	CA	5.2–6.4(30)(0.15% NaCl)	
	-1200				20.8–24.8	
	-3100		96–98%		4.0–4.8	
	-3200				16.0–18.4	

TABLE XVII-1 (*Continued*)

List of Membrane Manufacturers in Japan (1984)

Manufacturer	Trade Name	Shape	Fractional/ Rejec-Mole. Wt./ tion (%)	Material	Flux $(1/m^2hr)(kg/cm^2)$
	-5100		94–96%(3.5% NaCl)		1.8–2.0(56)(25°C, 3.5% NaCl)
	-5200				7.2–8.0
	SP-110		99.5%	PET	1.6
	-120				6.5
Toyobo Co. Ltd.	*RO*				
	HR-3110	Hollow Fiber	90%(0.05% NaCl)	CA	0.3(10)(25°C, 0.05% NaCl)
	-5110				1.0
	-5230		90%(0.15% NaCl)		9.0(30)(25°C, 0.15% NaCl)
	-5330				15.0
	-8330				40.0
	-5155		99%(3.5% NaCl)		1.0(55)(25°C, 3.5% NaCl)
	-5355				3.5
	-8355				10.0
Toyo Filter Co. Ltd.	*UF*				
	UF-10	Flat Sheet	10,000	PS	180–300(4)(25°C, PW)
	-50		50,000		600–1,800
	-200		200,000		600–1,800(1)
	UP-20		20,000	PA	360–600(4)
	UH-05		500	PEC	12–24
	-1		1,000		18–30
Yuasa Battery Co. Ltd.	*UF*				
	ED-01	Flat Sheet	64,500	PVC	15,600(1)(25°C, PW)
	-02	Tubular & Spiral	65,000		12,000
Mitsubishi Rayon	*RO*				
	4014	Spiral	97.5%(0.1% NaCl)	CA	27(28)(25°C, 0.1% NaCl)

Engineering Co. Ltd.	4018	$(8.3m^2)$	97%		34
	4022		96%		41
	B441		99%	PA	30
	B451		98.5%		38
	B461LP		96%		38(17)
	BW30HP-4040	Spiral	98%	PS	27(28)
	SW30-4040	$(6.5m^2)$	98.6% (Sea Water)		31(56)(25°C, Sea Water)
	MRG-5	Spiral	50% (0.1% NaCl)		30(20)(25°C, 0.1% NaCl)
	-10	$(5.1m^2)$	30%		48(14)
	-20		15%		45(7)
	-50		10%		76(7)
	UF				
	MRT-104	Tubular	13,000	PAN	75(1)(25°C, PW)
	MRE-500U60	Spiral	60,000	PS	70
	MRF-500U60	$(5.1m^2)$	60,000	NCA	85
	MRS-15	Tubular	15,000	PS	180
	-60		60,000		230
	-100		100,000		350
	-150		150,000		450
	PKO-614A	Cartridge	150,000	PE	234
IHI Co. Ltd.	*RO*				
	R-1	Tubular (Outside)	98% (0.5% NaCl)	CA	10(42)(20°C, 0.5% NaCl)
	-2		95%		30
	-3		90%		40
	-4		80%		59
	H-1		90% (2% Sucrose)		50(28)
	-2		80%		75
	-3		60%		100

TABLE XVII-1 (*Continued*)

List of Membrane Manufacturers in Japan (1984)

Manufacturer	Trade Name	Shape	Fractional/ Rejec-Mole. Wt./ tion (%)	Material	Flux $(1/m^2hr)(kg/cm^2)$
	UF				
	F-1		6,000		
	-2		20,000		
	-3		35,000		
TDK	*UF*				
Co. Ltd.	DC-0505	Tubular (Outside)	100,000	Porous Ceramics	120(1)(25°C, PW)
Ulvac	*UF*				
Service Co.	G-05H	Flat Sheet	500	Poly-	3(1)(25°C, PW)
Ltd.	-10H		1,000	Styrene	7
	-05T		5,000		14
	-10T		10,000		22
	A-15T		15,000	CA	30
	-50T		50,000		100
	S-15T		15,000	PS	100
	F-50T		50,000	VDF	100

PAN: Polyacrylonitrile, PS: Polysulfone, PO: Polyolefine, PES: Polyethersulfone, PBIL: Polybenzimidazolone, PI: Polyimide, PVA: Polyvinylalcohol, PET: Polyether, NCA: Non-cellulosic, PE: Polyethylene, VDF: Vinylidene fluoride, PW: Pure Water

REFERENCES

1. T. Doi, "A Report for Scientific and Industrial Membrane Development in Japan" paper presented at Europe–Japan Joint Congress on Membranes and Membrane Processes, Stresa, Italy, June 1984
2. H. Yoshitome, "Gas Membrane Separation in C1 Chemistry Project" ibid
3. H. Yoshitome, "R & D of Membrane Technologies in Ministry of Agriculture, Forestry and Fishery" ibid
4. H. Yoshitome, "Membrane Technology in Water Re-Use Promotion Center" ibid
5. T. Nakagawa, "Research of Membrane in Japanese Universities under the Ministry of Education" ibid

Investigation of Structure-Permeability Relationships in Silicone Polymers

MARTY CIFUENTES, C.L. LEE, H. CHAPMAN, K. LEE, AND K. ULMAN

Dow Corning Corporation, Midland, Michigan

The effect of polymer structure on the gas transport properties of polysiloxane polymers has been investigated. For the $(MeRSiO)_n$ series of polymers, the oxygen and nitrogen permeability coefficient was found to be dependent on polymer Tg. The relationship between selectivity and permeability for $(MeRSiO)_n$ will also be compared with those of organic polymers.

Dow Corning and Syracuse University are currently involved in a contract designed to investigate relationships between polymer structure and the resultant gas permeability behavior observed for a variety of polysiloxane materials. Using polydimethylsiloxane as a benchmark, silicone polymeric structure is systematically modified at either the alkyl side group position or within the polymer chain backbone itself. These types of substitutions create changes in the physical and conformational properties of polymers. This, in turn, affects transport characteristics.

All of the synthesized polymers are first characterized. Oxygen and nitrogen permeability coefficients are then measured for thin, vulcanized membranes. The apparatus used to measure gas permeability coefficients is the variable volume permeability cell developed by Professor S.A. Stern. Samples are then sent to Syracuse University where more extensive testing is carried out for a variety of commercially important gases under varied temperature and pressure conditions. Syracuse also has the capability of directly measuring the gas solubility in polymeric materials.

The work presented today was generated during the first year of the contract and is restricted primarily to the $(MeRSiO)_x$ series of polymers. The behavior exhibited by this family of polymers is typical of that observed for other polymers having different types of structural modification.

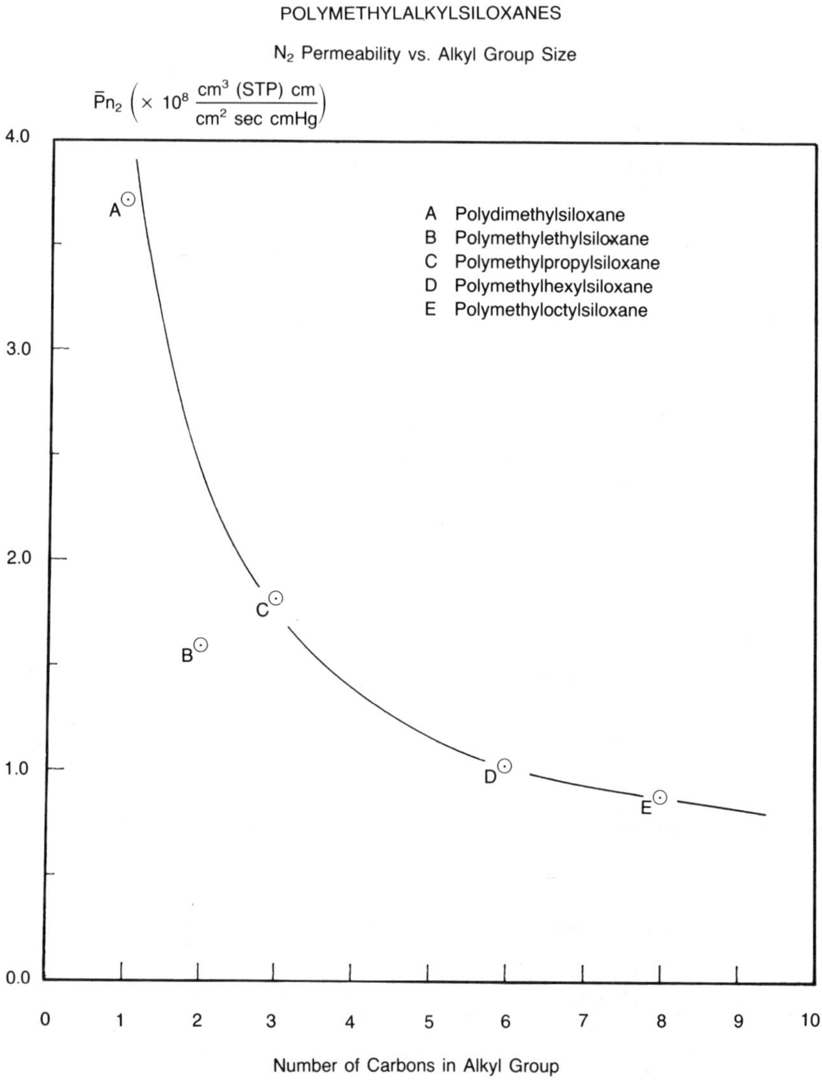

FIGURE 1

Nitrogen permeability coefficients ($\bar{P}N_2$) were plotted for the $(MeRSiO)_x$ series of polymers as a function of increasing size of the substituted R group. In general, gas permeability steadily decreased with increasing size of R (Figure 1). However, polymethylethylsiloxane, $(MeEtSiO)_x$,

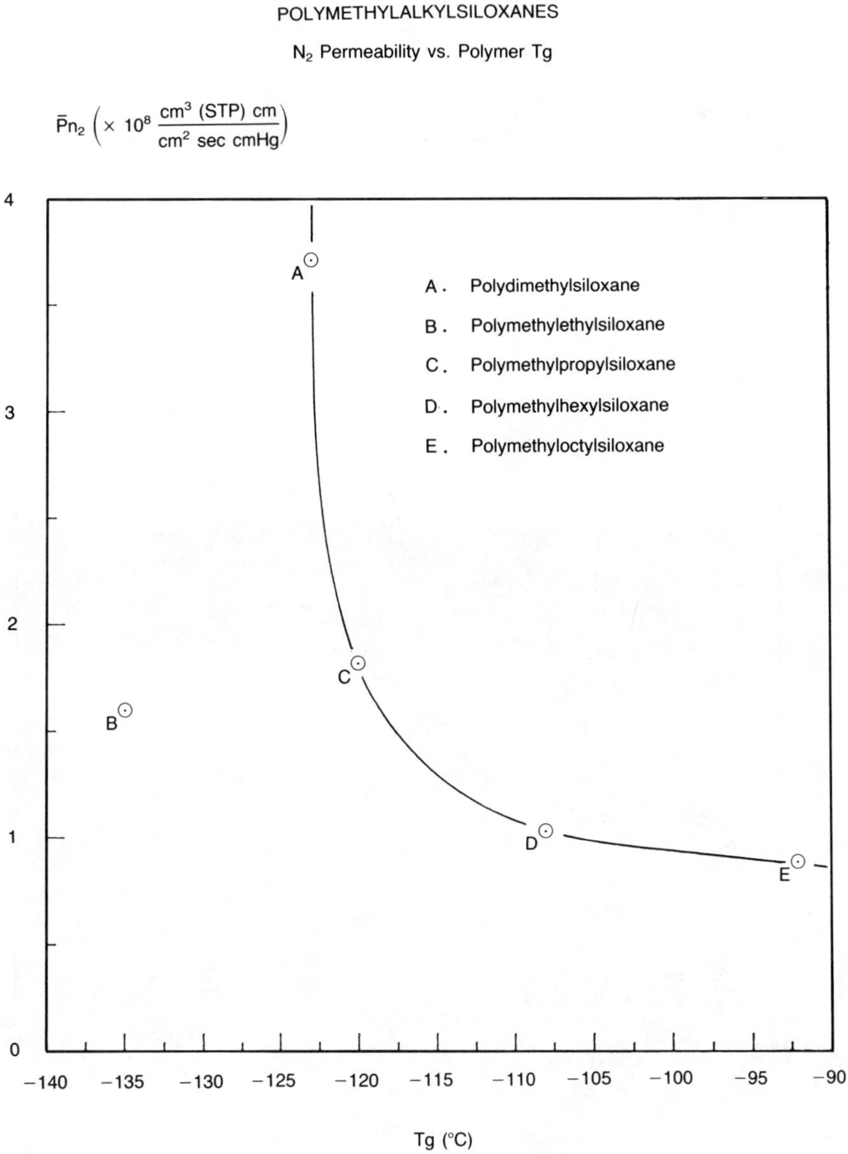
POLYMETHYLALKYLSILOXANES
N2 Permeability vs. Polymer Tg
P̄n2 (× 10^8 cm^3 (STP) cm / cm^2 sec cmHg)
4
3
2
1
0
A
B
C
D
E
A. Polydimethylsiloxane
B. Polymethylethylsiloxane
C. Polymethylpropylsiloxane
D. Polymethylhexylsiloxane
E. Polymethyloctylsiloxane
−140
−135
−130
−125
−120
−115
−110
−105
−100
−95
−90
Tg (°C)

FIGURE 2

N_2 PERMEABILITY Vs. O_2/N_2 SELECTIVITY

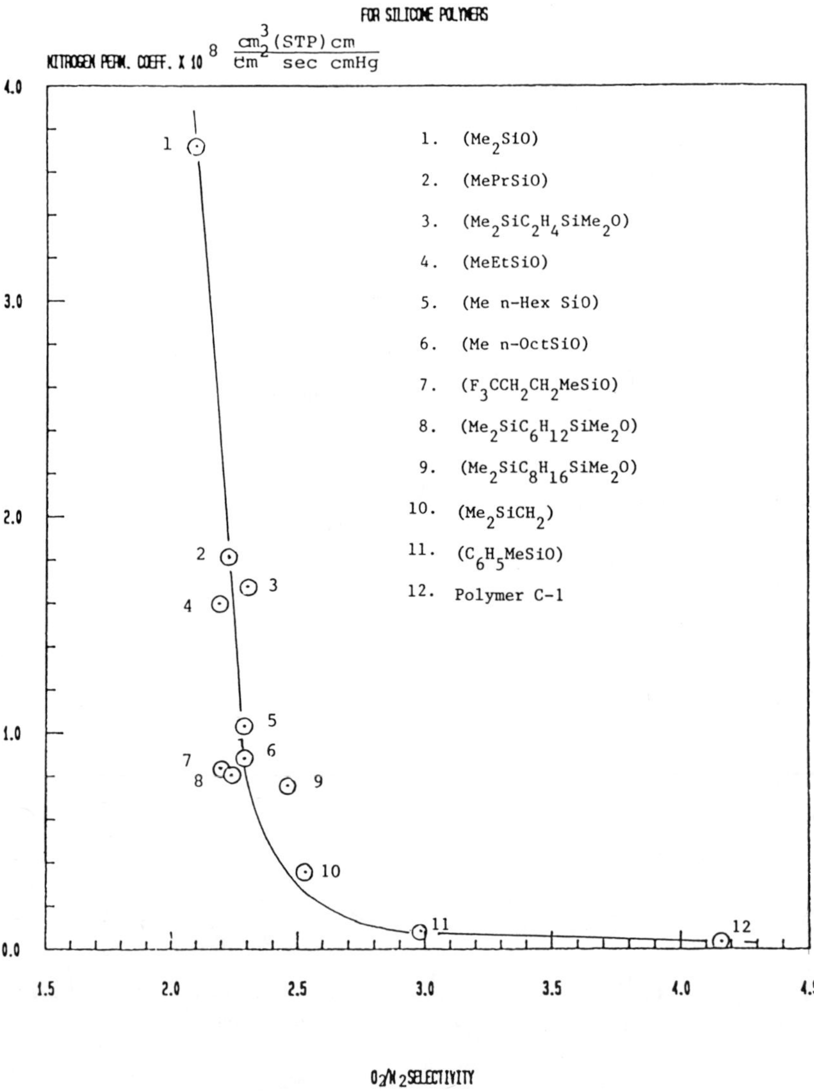

FIGURE 3

deviated from an otherwise smooth trend by exhibiting lower gas permeability than polymethylpropylsiloxane, $(MePrSiO)_x$. The anomalous behavior of $(MeEtSiO)_x$ is illustrated even more dramatically when $\bar{P}N_2$ is plotted as a function of polymer glass transition temperature (Tg). One would expect gas permeability to decrease as polymer T_g increases. In general, this behavior is indeed observed (Figure 2). However, $(MeEtSiO)_x$ is only half as permeable as polydimethylsiloxane, although it has a lower Tg (−135°C).

Also included in today's presentation is a graph which plots the permeability-selectivity relationships for all polysiloxane materials that have been evaluated to date. One of the most well-known transport relationships established for organic polymers is the observed trade-off between permeability and selectivity. This trend is also exhibited in the case of silicone polymers, although the polymers listed represent a variety of structural modifications (Figure 3). These changes include size differences of the attached alkyl side group, the incorporation of hydrocarbon linkages into the siloxane chain backbone, and the introduction of polarity into the polymer.

Work in the area of gas separations is still ongoing. It is hoped that the results from our studies will enable us to better predict transport properties based on polymeric structure, and teach us the most efficient way to chemically tailor membranes in order to achieve a desired separation in a defined commercial application.

ACKNOWLEDGEMENTS

This work was sponsored jointly by the Gas Research Institute and Dow Corning Corporation.

Effect of Ionizing Radiation on Ethylene Vinyl Acetate Copolymer Hollow Fibers

A. KEITH FRITZSCHE

Monsanto, St. Louis, Missouri

The effects of irradiation induced crosslinking on ethylene vinyl acetate copolymer hollow fibers were studied. The irradiation induced crosslinking was found to improve the thermal and solvent resistance of the treated fiber. The irradiation caused the physical properties such as yield stress and elastic modulus to pass through maxima with increasing dosages. Irradiation was also seen to be more beneficial if oxygen was not present during the treatment. Finally, unlike the low density polyethylene fibers studied, irradiation induced crosslinking reduced the phenol transport rates through the ethylene vinyl acetate copolymer hollow fibers.

INTRODUCTION

Hollow fiber membrane systems from ethylene copolymers were developed to remove phenol from aqueous industrial waste streams by reactive dialysis (1). Although these fibers demonstrate good phenol permeation rates in pure aqueous phenol feeds, industrial waste streams generally contain a variety of components. Such absorption swells the fibers resulting in loss of physical strength and in many cases failure by rupture.

Crosslinking is known to improve the solvent resistance of polymeric materials. One method commonly used to induce crosslinking in many types of polymers is ionizing radiation. R.J. Lee, for instance, irradiated polyethylene membranes and found that irradiation induces crosslinking of the polymer, improving its temperature stability and resistance to solvents (2). G. Zeppenfeld showed similar results with polyethylene fibers, and he observed that the treated fibers also possessed improved strength and aging characteristics as well as reduced shrinkage (3). E.V. Kirkland and J.F. Jennings noted that polyethylene films have improved permeation properties if, prior to use, they were irradiated (4). S. Yamada and T. Takeo observed that the permeation rates of benzene during re-vaporation through irradiated polyethylene membranes were unchanged at dosage levels less than 50 Mrads, but that the permeation of benzene exhibited a marked decrease in rate at higher dosages (5). These investigations also reported that irradiation crosslinking at dosages at 10 Mrads had little effect on either the permeation rates or separation factors of membranes made from poly (4-vinylpyridine) and polystyrene in spite of the change in the degree of swelling and flexibility of the polymers (6). M.G. Katz and T. Wydeven Jr. used ionizing radiation to prepare crosslinked polyvinylalcohol membranes in order to study their water and salt transport properties (7). Irradiation crosslinking has also been used to prepare ion exchange membranes of polyacrylic acid and polyacrylate salts (8).

Since polyvinyl acetate and ethylene vinyl acetate copolymers undergo radiation induced crosslinking (9-11), it was felt that the improved thermal and solvent stability, seen with the radiation crosslinked polyethylene films and fibers, could also be incorporated into the ethylene vinyl acetate copolymer hollow fibers using a similar technique. For this reason, hollow fibers spun from polyethylene and ethylene vinyl acetate copolymers were crosslinked by irradiation. The treated fibers were then studied as to the changes induced in their temperature and solvent resistances with increasing degrees of crosslinking.

EXPERIMENTAL

Sample Preparation

Hollow fibers were melt extruded from low density polyethylene (Rexene PE 126, Melt Index = 30), 12% vinyl acetate copolymer of ethylene (DuPont Alathon 3130, MI = 2.5), and 18% vinyl acetate copolymer with ethylene (DuPont Alathon 4227, MI = 18). The fibers were wound on bobbins to a depth of 1/2 inch. The bobbins, one for each test condition, were shipped to Radiation Dynamics, Inc. of Westbury, New York for irradiation using a 3 MeV Dynamitron to generate the electron beam. The bobbins were positioned horizontally on a cart which was passed through the electron beams. During each pass the bobbin received a dosage of 2.5 Mrads. After each pass the bobbin was rotated 180° so that the lower surface of the bobbin would be exposed directly to the beam. Each pass required 24 seconds, and the interval between each pass was sufficient to dissipate any heat generated in the fibers during exposure to the electron beam. The bobbins which were to be irradiated in a nitrogen atmosphere were placed in polyethylene bags. Nitrogen was flushed through the bags to remove the air and subsequently the bags were heat sealed.

Permeation Measurements

The phenol permeability of each fiber was determined by assembling a skein of 60 fibers in a loop through a sealed rubber stopper and providing the respective ends with a nipple to pass fluids through the fiber bores. The looped skein of fibers was immersed in a stirred constant temperature bath containing an aqueous phenol solution. A 15% sodium hydroxide solution was passed through the fiber bores to convert the permeating phenol to the nonpermeable sodium phenate. The existing bore fluid was collected, diluted to a given volume, and the concentration of the sodium phenate present in the effluent determined by absorption of ultraviolet light at 287 mμ. The amount of phenol permeated through the hollow fibers was calculated from the concentration of the sodium phenate and the time of the test, generally one hour.

The permeation rates of phenol through these hollow fibers were measured at 70°C. The results are shown in Table I for fibers spun from low density polyethylene (LDPE), 12% vinyl acetate copolymer with ethylene (12% EVA), and 18% vinyl acetate copolymer with ethylene (18% EVA). Identical permeabilities were obtained from fibers taken at any depth in the bobbin. This indicated that the electron beam penetrated the bobbin forming a uniform number of crosslinks throughout the bundle. The tested fibers were irradiated in the air with dosages from 0 to 100 Mrads. Only the LDPE hollow fiber irradiated at the maximum level could be tested, however. The two EVA hollow fibers exposed to an identical dosage were fused together and embrittled by the intensity of the electron beam. These fibers broke in numerous places and could not be removed from the bobbin. Nor could any data be obtained from the unirradiated 18% EVA hollow fibers. During the 70°C test, the fibers in this skein failed by rupture. The irradiated 18% EVA fibers maintained

TABLE I

Variation of Permeation Rate of Phenol with Increasing Radiation

	$P(70°)\frac{cm^3 - cm}{sec - cm^2 - \Delta V} \times 10^8$		
Dosage (Mrads-Air)	LDPE (0% VA)	EVA (12% VA)	EVA (18% VA)
0	2.5	90	—
10	2.6	54	66
20	3.2	31	13
40	3.5	25	21
100	4.6	—	—

their integrity, but were tacky and tended to stick together. This tackiness accounts for the spurious permeability results also reported for the 18% EVA hollow fibers at 70°C.

Definite trends were observed in the LDPE and the 12% EVA hollow fibers. As the applied electron beam dosage increased, the permeabilities of the LDPE hollow fibers increased while those of the 12% EVA hollow fibers decreased.

The increased permeability of the LDPE hollow fibers can not be readily explained. Radiation can change the permeability of a polymer by disrupting the crystallites, forming crosslinks between the polymer chains, inducing chain scissions, or altering the chemical structure.

Slichter and Mandel showed that at high dosages irradiation inflicts changes concurrently in the crystalline and amorphous regions of polyethylene (12). Lattice defects are induced in the crystallites, and the characteristic separations between chains are altered in both the crystalline and amorphous regions. A reduction in crystallinity would reduce the impermeable regions in the hollow fiber increasing the solubility of the phenol in the membrane and decreasing the tortuosity of the diffusion path. The crystalline regions also act as crosslinks which restrict the mobilities of neighboring chain segments. Reduction of the crystallinity is expected to yield enhanced chain mobility and enhanced permeability. Charlesby argues that this is the case for even low radiation ranges such as those used here (13). However, Bixler, Michaels, and Salame showed that little destruction of crystallinity would occur in this radiation range (14). In fact, they showed that the change in density of the polymer after irradiation appears to stem simply from a densification of the amorphous phase with no attendant reduction in crystallinity or instability of the crystallites. Yet, both Bent (15) and Chmutov (11) observed increases in permeability constants for organic liquids and water in polyethylene irradiated at these levels. Therefore, it seems likely that the enhanced permeability of the LDPE hollow fibers results from either chain scission or increased polarity of the polymer due to the incorporation of the oxygen.

In the EVA hollow fibers, the results differed. These polymers have a lower degree of crystallinity than does LDPE. The crosslinking occurs predominantly, if not entirely, in the amorphous region. E.J. Lawton, J.S. Balwit, and R.S. Powell showed that at an equivalent dose irradiation is more effective in producing crosslinks when the material is irradiated in a completely amorphous state than when it is irradiated in a partly crystalline state (17). In the EVA system, more crosslinks are introduced into the hollow fiber than in the LDPE system at the same irradiation dose. In the copolymer case, reduction in chain mobility by formation of crosslinks in the amorphous regions would exceed any contribution to enhanced permeability due to chain scission or the incorporation of oxygen groups into the polymer. Therefore, lower permeation rates would result. In fact, removal of the polar acetate groups from the polymer by chain scission would further reduce the permeability.

Shrinkage

The shrinkage of irradiated fibers and untreated controls was determined by immersing skeins containing 20 fibers of measured length in a constant temperature water bath for one hour. The lengths of the fiber skeins were measured again after the heat treatment. The temperature of the bath was increased and after the bath reached constant temperature the test was repeated. The shrinkage was measured over a range of temperatures from 25 to 90°C for the 12% and 18% EVA hollow fibers. The results are given in Tables II and III. As the temperature was increased all of the fibers suffered shrinkage. Yet, the shrinkage progressively decreased with increasing dosage. Also, the temperature of shrinkage onset increased with increased irradiation. For instance, the unirradiated 12% and 18% EVA hollow fibers began to shrink at 60°C, but no shrinkage appeared to either the 12% or the 18% EVA irradiated with 40 Mrads under nitrogen until 80°C.

It was also observed that the degree of shrinkage in the fibers irradiated under nitrogen was less than that obtained when the fibers were irradiated in air at the same dosage. These results are shown graphically in Figures 1

TABLE II

Percent Shrinkage of 12% Ethylene Vinyl Acetate Hollow Fibers Irradiated in Air and Nitrogen

Temp. (°C)		% Shrinkage				
	(Dosage Mrads)	0	20A	20N	40A	40N
25	0	0	0	0	0	0
60		1.9	0	0	0	0
70		3.8	0.5	0	0	0
75		4.8	1.9	1.0	0	0
80		7.7	4.8	1.9	1.0	1.1
85		13.4	9.6	4.8	6.7	3.1
90		45.2[a]	30.8	19.2	21.9	18.7

[a] Fibers fused together.

TABLE III

Percent Shrinkage of 18% Ethylene Vinyl Acetate Hollow Fibers Irradiated in Air and Nitrogen

Temp. (°C)	% Shrinkage				
(Dosage Mrads)	0	20A	20N	40A	40N
25	0	0	0	0	0
60	3.8	0	0	0	0
70	7.7	1.9	1.9	1.0	0
75	11.5	4.8	3.8	3.8	0
80	21.0[a]	12.4[a]	11.5	9.6	5.9
85	42.0[a]	21.0[a]	21.2	14.4	9.8
90	Melted	22.0[a]	23.1[a]	15.4	9.8

[a] Fibers fused together.

and 2. Figure 1 shows the shrinkage of the 12% EVA hollow fiber irradiated in nitrogen as a function of temperature. Figure 2 compares the shrinkage of unirradiated 18% EVA hollow fibers with those irradiated with 40 Mrads dosage in air and nitrogen.

Solvent Extraction

The percentage of extractables would decrease with increasing radiation dosage provided that the rate of crosslinking is greater than the rate of chain scission. The amount of extractable components in the irradiated 12% EVA fiber was determined by placing weighed fibers into a Soxhlet extractor and exposing them to hot xylene for 48 hours. The fibers were then dried overnight in the hood. They were placed in an oven for 24 hours at 70°C and weighed. Finally, the fibers were placed in a vacuum desiccator for 24 hours and re-weighed. The results in Table IV show that the uncrosslinked samples totally dissolved in hot xylene and that the nitrogen irradiated fibers generally had a lower percentage of extractables than did the air irradiated samples.

These results on the percent extractables and those obtained from the shrinkage experiments indicate that the presence of oxygen during irradiation result in a reduction of the rate of crosslinking. These observations agree with those of Chapiro, who irradiated polyethylene in the presence of oxygen (18).

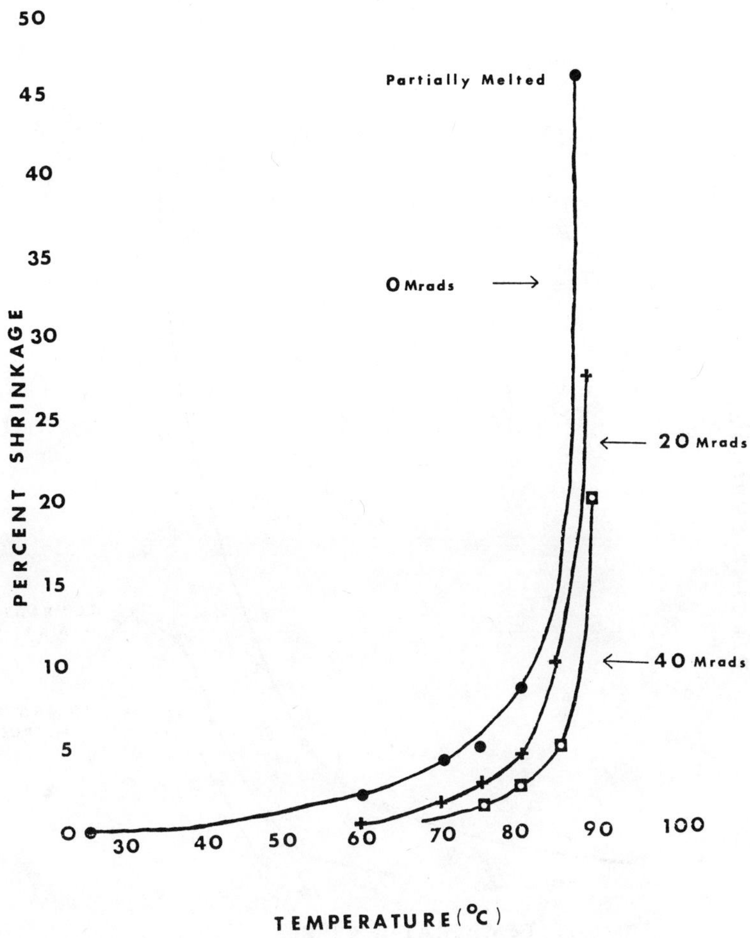

FIGURE 1 Shrinkage of 12% Ethylene Vinyl Acetate Irradiated in Nitrogen.

Physical Property Measurements

The physical properties of irradiated and unirradiated hollow fibers were measured by T. Murayama on the Rheo-Vibron at 11 cps. The yield stress and the elastic modulus of the fibers were measured in a 3% phenol solution at 70°C. The dynamic mechanical properties of the 12% EVA fibers in air were also measured as a function of temperature.

The physical strength measurements obtained in the 3% phenol solution are given in Table V. The results are similar to those reported by Vermes

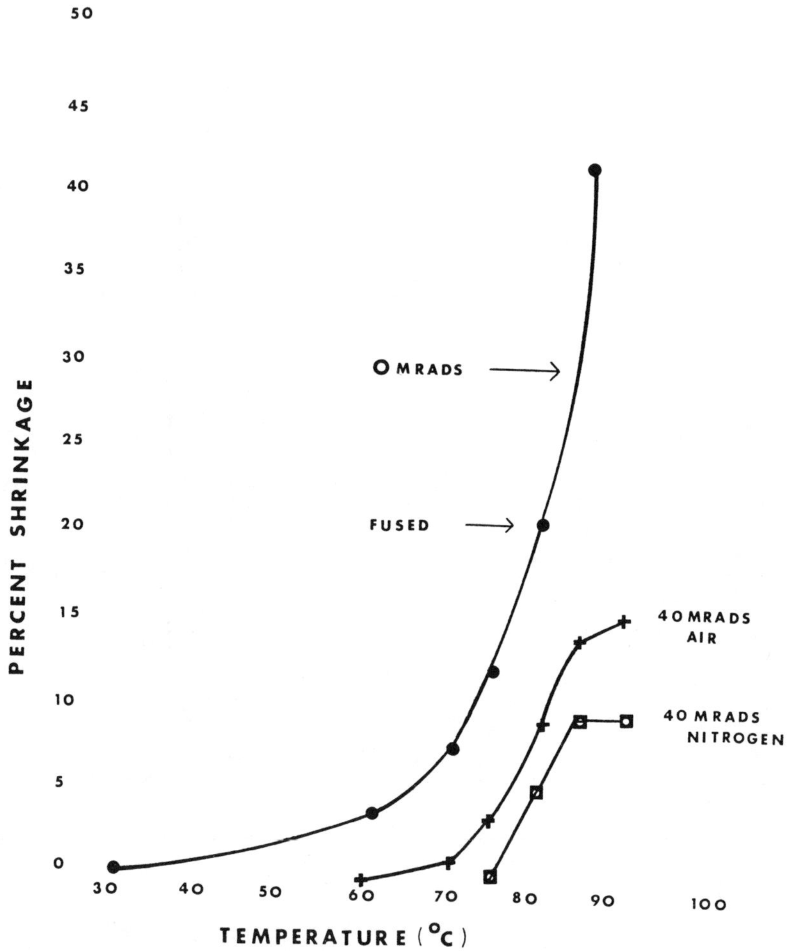

FIGURE 2 Comparison of Shrinkage of 18% Ethylene Vinyl Acetate Fibers Irradiated with 40 Mrads in Air and Nitrogen.

and Brenner (11). The mechanical properties reached a maximum at a given total ionizing dosage and began to decline upon further exposure. The largest yield stress and elastic modulus of the LDPE were observed at 40 Mrads dosage and a marked decline in these values as seen at 100 Mrads dosage. These results suggest that maxima in these values occurred between 40 and 100 Mrads for LDPE. The maxima in the 12% EVA hollow fibers occurred at lower values. The maximum in the yield stress was observed at 20 Mrads and declined upon further irradiation. A drop

TABLE IV

Percent Weight Loss of EVA Hollow Fiber Extracted with Xylene

Dosage	Percent Weight Loss Air Irradiated	Nitrogen-Irradiated
0	Dissolved	Dissolved
20	19.9	7.0
40	18.3	5.4
60	4.1	6.9
80	4.6	5.0

TABLE V

Physical Strength Measurements for Irradiated Hollow Fibers (3% Phenol, 70°C)

Material	Dose (Mrads)	Yield Stress (psi)—	$E \times 10^{-8}$ (dynes/cm^2)
LDPE	0	1000	5.9
LDPE	10	1200	6.7
LDPE	20	1200	7.6
LDPE	40	1500	7.9
LDPE	100	700	5.0
EVA (12% VA)	0	1070	3.1
EVA (12% VA)	10	1100	2.2
EVA (12% VA)	20	1600	3.1
EVA (12% VA)	40	1200	2.4

also was observed in the elastic modulus in this range. The validity of this particular measurement is questionable, however, due to the low value of the elastic modulus for the 12% EVA hollow fiber at 10 Mrads dosages.

Typical results for the variation of the dynamic mechanical properties of the 12% EVA hollow fibers in air as a function of temperature are shown in Figures 3 and 4. Figure 3 shows the dynamic mechanical properties of an unirradiated 12% vinyl acetate EVA, and Figure 4 shows the results for an identical fiber which was irradiated with a 40 Mrad dosage in air. Several differences were observed between the irradiated fiber and the control. The dynamic modulus (E′) of the irradiated fiber was less temperature dependent than the dynamic modulus of the control. The onset of the terminal zone, caused by the melting of the crystallites in the fiber and rearrangement of the chain conformations occurred at a lower temperature (75°C) in the unirradiated EVA fiber than in the

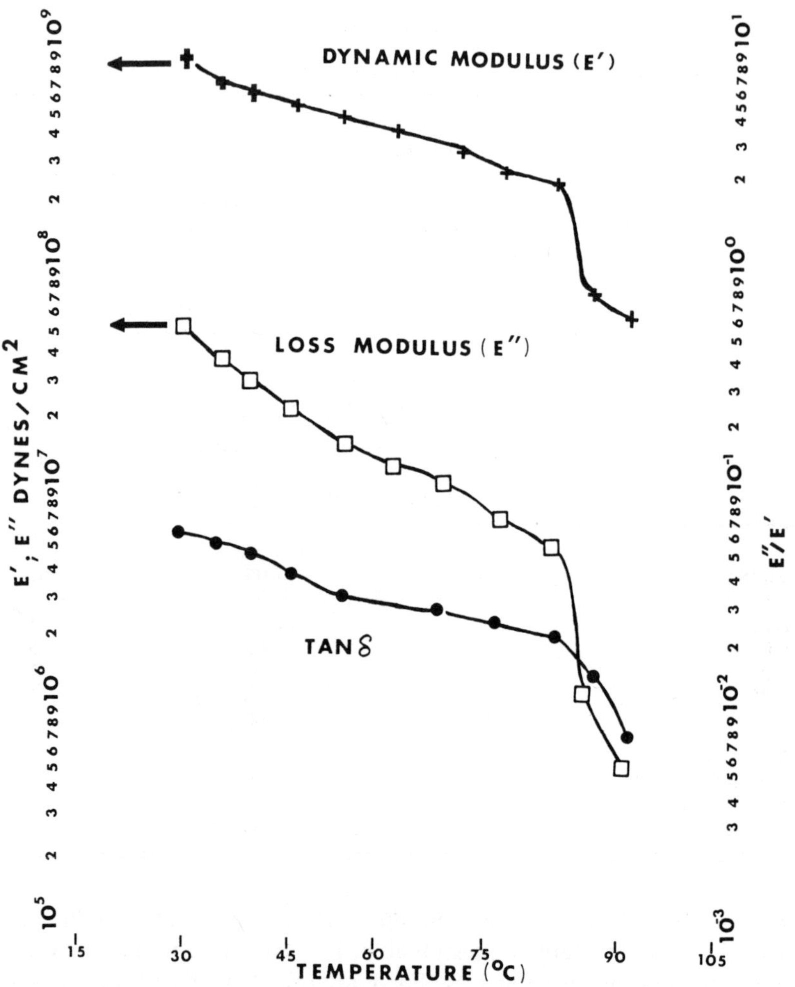

FIGURE 3 Dynamic Mechanical Properties of Unirradiated 12% Vinyl Acetate EVA Hollow Fibers (Vibron, 11 cps).

irradiated (82–83°C). A pseudoequilibrium zone appeared in the dynamic modulus of the crosslinked fibers above the melting region of the crystallites. This plateau is similar to that found in any crosslinked system above its melting point. The terminal region in the uncrosslinked fiber represents the unraveling of the chain entanglements and ultimate flow of the material. Finally, the dynamic and loss moduli were slightly lower below the melting point in the crosslinked fibers in contrast to the control. This behavior reflects disruption of some of the crystallites in the

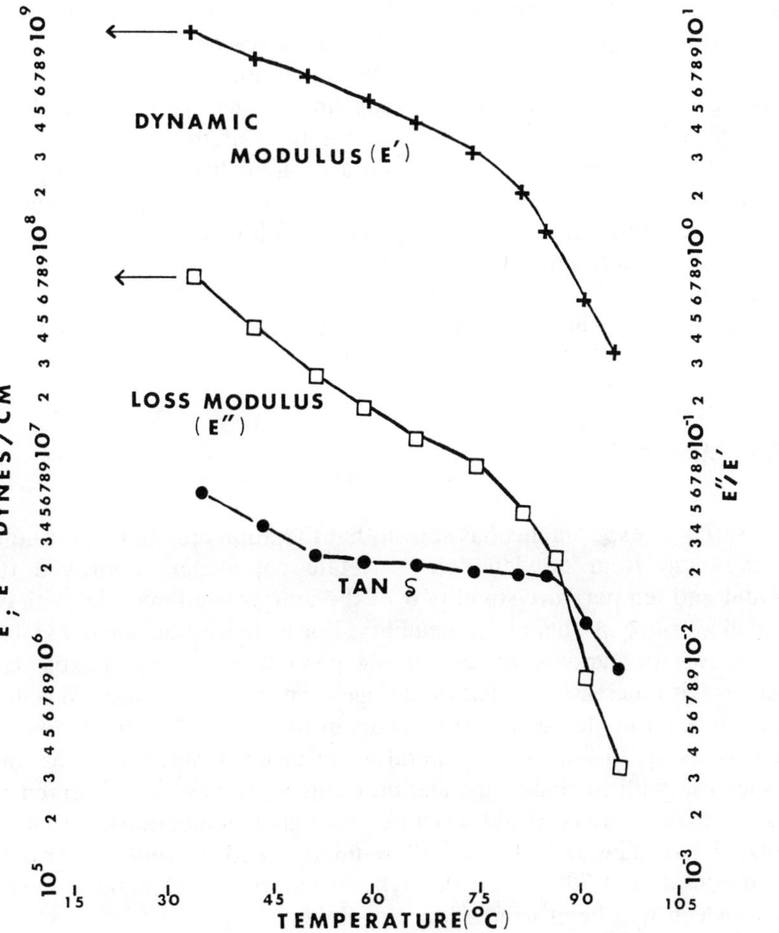

FIGURE 4 Dynamic Mechanical Properties of Irradiated (40 Mrads-Air) 12% Vinyl Acetate EVA Hollow Fibers.

fiber. It also implies that a degree of chain scission as well as crosslinking occurred.

Improved Fiber Lifetimes

Irradiated and nonirradiated 12% vinylacetate EVA hollow fibers were skeined into test cells using techniques described elsewhere (1). These cells were then placed in a simulated aqueous phenol feed saturated with organic constituents which were known to swell and weaken the fibers.

Failure would then occur by rupture formation. These ruptures could be detected by a rapid increase in pH of the feed reservoir. This pH increase resulted from pumping the caustic sweep solution from the fiber bores into the feed reservoir. The operating time which elapsed until this pH increase was observed was defined as the fiber lifetime.

At 60°C, the unirradiated fiber had a six hour lifetime when operated in the simulated feed. The lifetime increased to 73–90 hours by lowering the feed temperature to 50°C and to 449–493 hours at 40°C. However, a dramatic improvement occurred when irradiated 12% vinyl acetate EVA hollow fibers (40 Mrads in nitrogen) were exposed to the same test conditions. These fibers survived between 81 and 125 days of continuous operation before failures were detected.

CONCLUSIONS

This series of experiments has shown that radiation crosslinking of hollow fibers made from ethylene vinyl acetate copolymers improved their solvent and temperature stability over the unirradiated material with only a small sacrifice in phenol permeability. For instance, the solvent stability was shown to increase by decreasing percentages of extractable components with increasing radiation dosages. The increased thermal stability was exhibited by decreasing shrinkages in the crosslinked fibers over the uncrosslinked. In fact, the temperature of onset of shrinkage was found to increase with increased irradiation dosages. It was also observed that the irradiated fibers could operate at higher temperatures than the unirradiated. The 18% EVA hollow fibers failed by rupture formation when operated at 70°C. Yet, this type of failure was eliminated in those fibers which had been irradiation crosslinked.

Irradiation also caused an initial increase in the physical properties at low density polyethylene and EVA hollow fibers. Maxima were observed in the yield streesses and elastic moduli of both systems. However, the maxima occurred at a much higher degree of irradiation dosage in polyethylene (40–100 Mrads) than in the 12% EVA system (10–20 Mrads).

Finally, it was observed that superior results were obtained when the EVA fibers were irradiated without oxygen present. Oxygen reduces the rate of crosslinking in these systems.

REFERENCES

1. A.K. Fritzsche and R.L. Leonard, U.S. Patent 4,082,658 (1978).
2. R.J. Lee, U.S. Patent 2,984,623 (1961).
3. G. Zeppenfeld, U.S. Patent 3,783,115 (1973).
4. E.V. Kirkland and J.F. Jennings, U.S. Patent 3,225,107 (1965).
5. S. Yamada and T. Takeo, *Kobunshi Ronbunshu*, 39, 23 (1982).
6. S. Yamada and T. Takeo, *Modu*, 6, 481 (1981).
7. M.G. Katz and T. Wydeven, Jr., *J. Appl. Polym. Sci.*, 26, 2935 (1981).
8. W.H. Philipp, C.E. May, and L.C. Hsu, Sci. Tech, *Aerosp. Rep*, 19 (1981).
9. K. Yonetani and W.W. Graessley, *Polymer*, 11, 222 (1970).
10. R.K. Tubbs, U.S. Patent 3,734,843 (1973).
11. R. Vermes and W. Brenner, "Irradiation of Polymers," p. 156, *American Chemical Society*, Washington, D.C. (1967).
12. W.P. Slichter and E.R. Mandel, *J. Phys. Chem.*, 62, 334 (1958).
13. A. Charlesby, "Atomic Radiation and Polymers," *Pergamon Press*, London (1960).
14. H.J. Bixler, H.S. Michaels, and M. Salame, *J. Polym. Sci*: Part A, 1, 895 (1963).
15. H.A. Bent, *J. Polym. Sci.*, 24, 387 (1957).
16. K. Chmutov and E. Finkel, *Zhur. Fiz. Khim.*, 33, 93 (1959).
17. E.J. Lawton, J.S. Balwit, and R.S. Powell, *J. Polym. Sci.*, 32, 257 (1958).
18. A. Chapiro, *J. Chim. Phys.*, 52, 246 (1955).

Commercially Successful Membrane Applications: Ion Analysis

TIMOTHY S. STEVENS

The Dow Chemical Company, Midland, Michigan

As originally developed (1), Ion Chromatography IC used as a suppressor a bed of ion-exchange resin called the "suppressor column". IC grew to be a most successful product, but as the years passed, complaints were made about several serious interferences associated with the suppressor column. Competitors made the most of these shortcomings, and their constant improvements threatened future prospects.

We solved many of the supressor column problems by developing a *hollow fiber suppressor* (2). The remaining problems were solved by developing an efficient means of *generating mixing* inside the hollow fiber, allowing much less membrane to do the same work (3). These developments advanced Ion Chromatography further ahead of the competition and have contributed to significantly increased world-wide sales by our licensee amounting to *thousands of units*.

This poster presentation will describe the development of the membrane suppression mode of Ion Chromatography. Include will be display units of membrane suppressors at various stages of development and complexity, a description of how mixing was generated inside the hollow fiber, and example chromatograms showing the superiority of the membrane suppressor.

REFERENCES

1. Small, Stevens, and Bauman, Analytical Chemistry, 1975, p. 1801.
2. Stevens, Davis, and Small, Analytical Chemistry, 1981, p. 1488.
3. Stevens, Jewett, and Bredeweg, Analytical Chemistry, 1982, p. 1206.

Commercially Successful Membrane Applications: The Hollow Fiber Suppressor for Ion Chromatography

TIMOTHY S. STEVENS

Dow Chemical Company USA, Michigan Division Analytical Laboratories, 574 Building, Midland, MI 48640

1984 MMI International Symposium on Synthetic Membranes, August 1984, Poster Session, See P18 0F Abstracts

Eluent Suppression Ion Chromatography is an important field of analytical chemistry, and the best performing suppressors to date are fabricated using ion exchange hollow fibers packed with plastic beads. The ion exchange hollow fiber allows continuous use of an ion chromatograph and eliminates interferences associated with the previously used suppresor column filled with ion-exchange resin. The beads generate mixing inside the hollow fiber which significantly reduces concentration polarization. The result is less membrane needed and a reduction of bandspreading to the point that it is insignificant. The commercial success of this development is indicated by the thousands of units sold in the past three years.

WHAT IS ION CHROMATOGRAPHY?

Most scientists and engineers working in the field of membrane science have probably never heard of Ion Chromatography and are not aware of the recent impact of membranes on this technique. Ion Chromatography is one of the most rapidly growing fields of analytical chemistry. Hundreds of applications have been published in areas as diverse as FBI investigations of terrorist bombings (1), to the analysis of beer (2). The first publication on Ion Chromatography appeared in 1975 (3).

The analytical power available for the determination of complex mixtures of many *organic* compounds was well developed in 1974 via gas and liquid chromatography and especially with gas chromatography-mass spectroscopy. This was not the case for complex mixtures of common *inorganic ions*, especially anions, despite the great need and high frequency of determination. Their analysis usually required separate wet chemical determinations or at best the use of ion-selective electrodes (if available) or separate atomic absorbance determinations of specific cations. The application of the well developed technique of *ion-exchange chromatography* to the problem was frustrated by the lack of a sensitive and reliable detector to place after the ion-exchange column.

The detector of choice for *general* use seemed to be the flow through an electrical conductivity cell since ions in solution are by definition conductive. However, ion-exchange chromatography also required an electrolyte in the eluent, and the conductivity detector responded to this electrolyte, significantly reducing the sensitivity of detection for the separated sample ions.

This problem was solved by placing a "suppressor column" filled with ion-exchange resin between the analytical column and the detector. The suppressor column reduced the conductivity of the specially selected eluent electrolyte and at the same time generally increased the conductivity of the separated sample ions.

In the example of Figure 1, injected anions are separated by ion-exchange chromatography using an eluent of dilute sodium carbonate. The suppressor column was filled with acid form ion-exchange resin, and it converted the highly conducting sodium carbonate to weakly conducting carbonic acid. The suppressor column also converted the sample ions to their acid moiety, and many common anions exhibit about five times higher conductivity as acids vs. their salts. Some anions exhibit minimal conductivity in their acid moiety, i.e. weak acid anions, and they are not ammenable to determination. As a matter of fact, weak acid anions are candidates for use as eluent electrolytes. Figure 2 is an example chromatogram showing the separation of common anions in commercial 50% NaOH produced by the diaphragm cell process.

COMMERCIAL SUCCESS BRINGS COMPETITION

The double-acting effect of the suppressor in reducing detection background signal while enhancing the signal from the separated sample ions is at the heart of the success of the technique. In 1975, the Dionex Corporation commercialized Ion Chromatography via a license from the Dow Chemical Company. The effective marketing of Ion Chromatography resulted in rapid growth for Dionex. As technological historians would have predicted, competitive techniques were soon developed, and the short-comings of the original technique were highlighted in the scientific literature (4) and in competitive advertising.

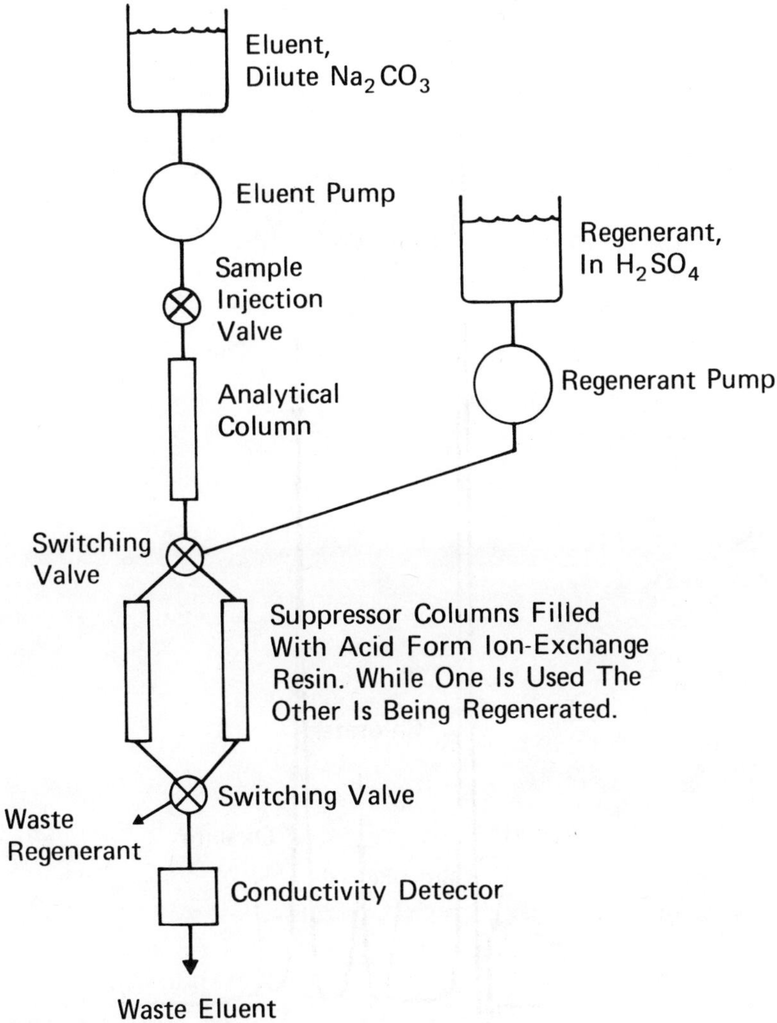

FIGURE 1 Ion chromatography using suppressor columns.

SUPPRESSOR PROBLEMS

The shortcomings of Ion Chromatography were problems with the suppressor column. It exhausted in use and required replacement or regenertion. Dionex Ion Chromatographs contained a regeneration pump, valves, and a controller for automated regeneration which kept

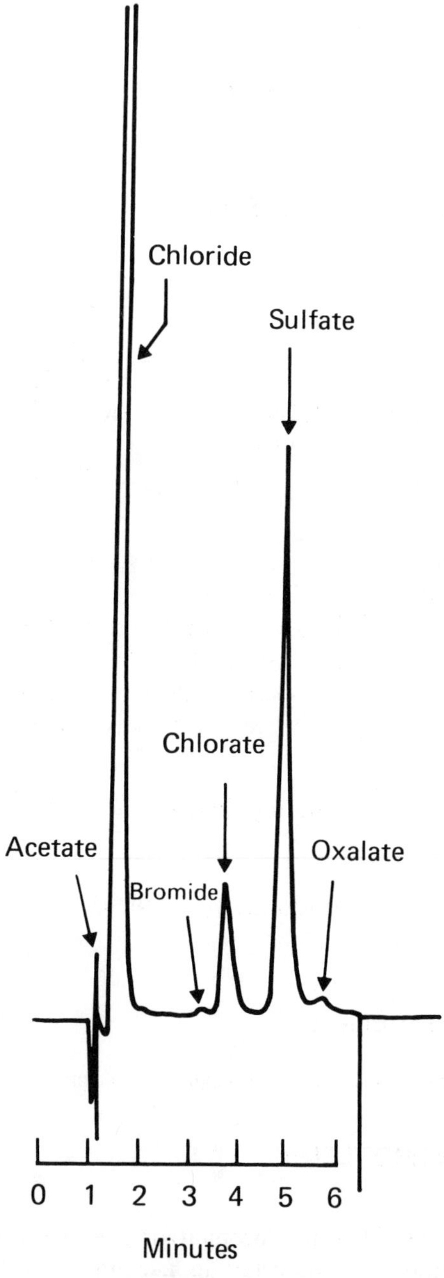

FIGURE 2 Determination of ions in 50% caustic using ion chromatography.

this complication to a minimum. More serious were interferences caused by the suppressor column with the determination of some ions. The suppressor column exhausts in use and presents the eluent stream with a continually decreasing volume of unexhausted ion-exchange resin. This resulted in varying elution times and/or varying responsiveness for some ions.

THE HOLLOW FIBER SUPPRESSOR

Figure 3 shows the theory of operation with a hollow fiber suppressor using an eluent of dilute sodium carbonate. The hollow fiber suppressor performs the same function as the suppressor column but can be operated at steady state by flowing a dilute solution of sulfuric acid around the outside of the hollow fiber. Therefore, the hollow fiber suppressor solves the problems of the use of an Ion Chromatograh, eliminates the need for the regeneration pump, valves, and controller, and does not result in varying elution times or response. However, the hollow fiber suppressor, at this stage of development, had one serious problem. In passing through the hollow fibers, the chromatographic peaks were broadened much more than with the use of suppressor column (5). Nevertheless, the Dionex Corporation was sufficiently impressed with the good qualities of the hollow fiber suppressor to take a license for commercialization.

NAFION HOLLOW FIBER

As we began to consider commercial development, various schemes were envisioned for producing the sulfonated polyethylene hollow fiber used in the first devices in large quantity. Then the use of commercially available Nafion 811-X ion exchange hollow fiber from DuPont was considered.

Nafion 811-X had a larger I.D. than desired (about 800 microns, wet) but was far stronger and more defect free than the previously used 400 micron ID sulfonated polyethylene hollow fiber. In the end, a single strand of Nafion 811-X, 20 feet long, was used in a device where sealing of the hollow fiber was accomplished by its own swelling in the close fitting bores of the inlet and outlet fittings, see Figure 4.

The resulting hollow fiber suppressor was rugged, reliable, and easily fabricated. However, it also showed just as much peak broadening as the devices using sulfonated polyethylene hollow fiber. It was envisioned that the device would find optional use in noncritical applications where con-

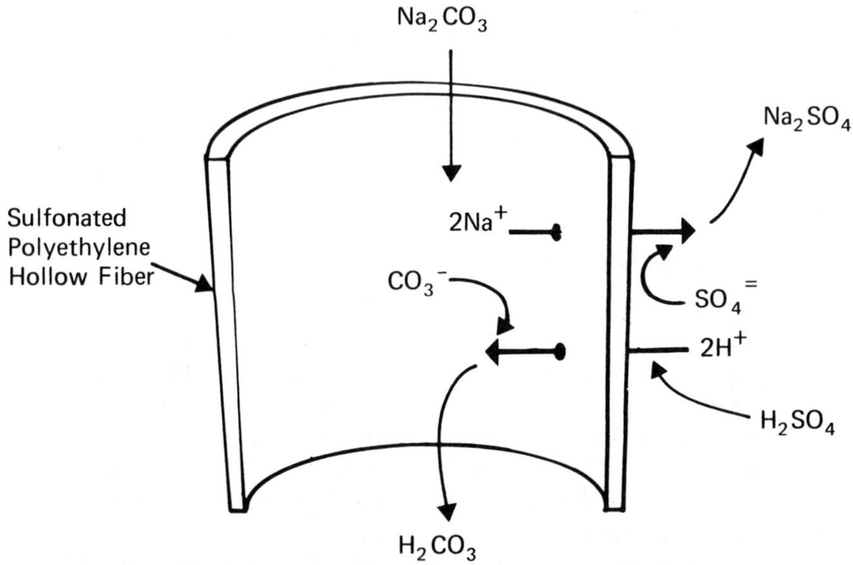

FIGURE 3 Theory of operation of the hollow fiber suppressor.

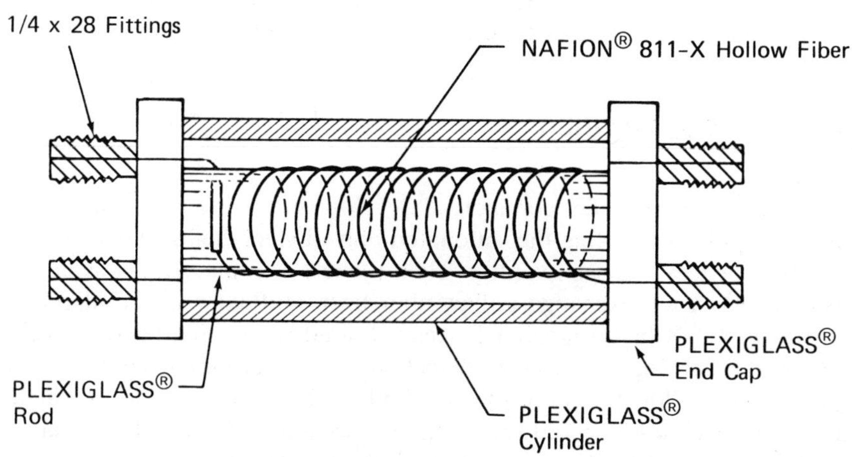

FIGURE 4 Fabrication of the hollow fiber suppressor.

tinuous operation and freedom from interferences were more important than peak broadening. However, the concurrent development of improved analytical columns giving even sharper peaks (6) made the problem of peak broadening even more acute.

THE PACKED HOLLOW FIBER SUPRESSOR

Experiments indicated that the resistance to mass transfer in the overall system was primarily in the bulk solution inside the hollow fiber and not through the membrane. A mixer was needed inside the hollow fiber to reduce concentration polarization. The solution to this problem was the placement of plastic beads inside the hollow fiber to promote mixing (7), see Figure 5.

This allowed a four-fold reduction in the length of hollow fiber needed (a 10-fold reduction if both sides of the membrane were packed with beads) and reduced the peak broadening problem so well that it became insignificant even with the new high performance analytical columns. A significant breakthrough had occurred at a most opportune time. Dionex

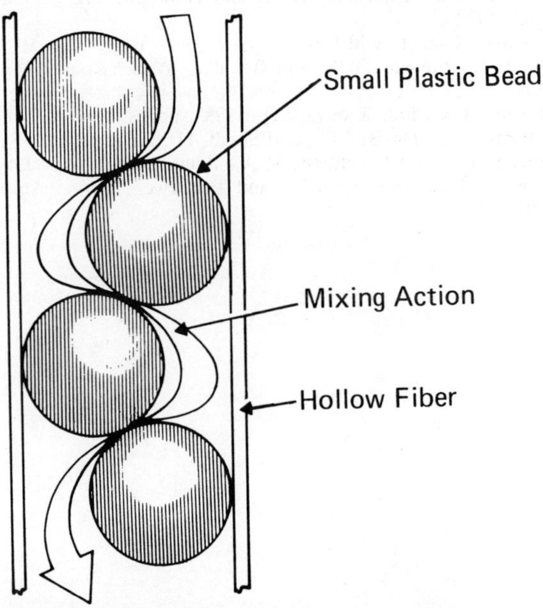

FIGURE 5 The packed hollow fiber suppressor.

had designed a new product line called the 2000 series Ion Chromatographs incorporating many improvements including the elimination of the regeneration pump, valves, and controller. With a 2000 series Ion Chromatograph the dilute sulfuric acid regenerant for the hollow fiber suppressor flowed simply by gravity. Now, the user of a Dionex Ion Chromatograph could enjoy the sensitivity and stability advantages of the suppressor mode of Ion Chromatography (a designation needed since competitive techniques also called themselves Ion Chromatography) without the previous problems of the use of a suppressor column. The instrumental and technological improvements offered by the 2000 series of instruments gained rapid acceptance from chromatographers and resulted in increased growth for Dionex. Thousands of membrane mode Ion Chromatographs have been sold in the last three years, while the demand for the suppressor column mode has almost disappeared. The hollow fiber suppressor for Ion Chromatography thus takes it place in the growing list of commercially successful membrane applications.

REFERENCES

1. Reutter, D.J., Buechele, R.C., and Rudolph, T.L., Anal. Chem., Dec. 1983, pages 1468–1472A.
2. Rocklin, R.D., Liquid Chrom., Nov. 1983, pages 504–507.
3. Small, H., Stevens, T.S., and Bauman, W.C., Anal. Chem., Nov. 1975, pages 1801–1809.
4. Stevens, T.S., Ind. R & D, Sept. 1983, pages 96–99.
5. Stevens, T.S., Davis, J.C., and Small, H., Anal. Chem., Sept. 1981, pages 1488–1492.
6. Stevens T.S. and Langhorst, M.A., Anal. Chem., May 1982, pages 950–953.
7. Stevens, T.S., Jewett, G.L., and Bredeweg, R.A., Anal. Chem., July 1982, pages 1206–1208.

Investigation of Progesterone Permeation Through Unfilled $[Me(C_nH_{2n+1})SiO]_x$ Membranes

K. L. ULMAN, C. L. LEE, H. L. CHAPMAN,

Silicone Research Depart., Dow Corning, Corporation, Midland, Michigan 48640

Flux rates and diffusion coefficients for progesterone through $(MeRSiO)_p$ (R = Me, Et, Pr, Hex, Oct) membranes, mounted in Ghannam-Chien membrane permeation cells, were quantitatively measured at 37°C. Boundary layer effects were minimized by adjusting the membrane thickness and applying the appropriate mathematical corrections.

Initially, the flux rate decreased as the size of the R group went from Me to Et; however, as the size of the R group increased from Et to Oct, the flux rates began to increase to the value obtained from R = Me. The diffusion coefficients, on the other hand, decreased consistently as the size of the R group increased.

In the fall of 1983, Dow Corning Corporation and Rutgers College of Pharmacy embarked on a joint program designed to establish a molecular model for predicting drug delivery rates through silicone polymers.

In this program, Dow Corning will investigate the effect of polymer structure on drug permeation rates — using progesterone and one other drug; whereas, Rutgers College of Pharmacy will investigate the effect of drug structure on drug permeation rates — using a couple of silicone polymers. Once all of the data are collected it will be used to develop a molecular model for predicting drug delivery rates through silicone polymers.

This paper presents the preliminary results obtained by Dow Corning on the release rates and diffusion coefficients of progresterone through methylalkylsiloxane membranes, where the alkyl substituents were Me, Et, Pr, Hex, and Oct.

In this work, a Ghannam — Chien membrane permeation apparatus was used (Figure 1). The apparatus consisted of two concentric cells which were separated by a membrane. A solution of 40% PEG 400 was used as the solvent and all of the permeation studies were conducted at 37° (to simulate body temperature).

The transport of progesterone through the membrane was monitored with a UV Spectrophotometer and the amount of drug that permeated

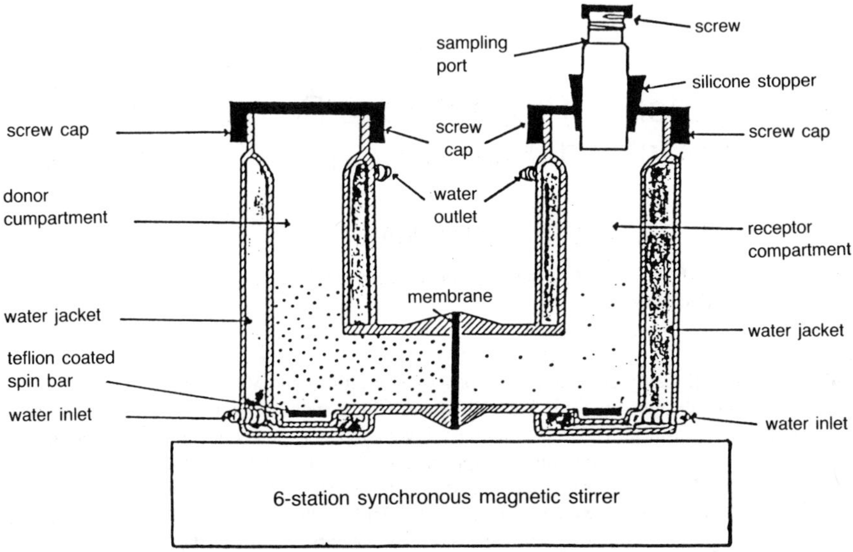

FIGURE 1 Ghannam-Chien membrane permeation apparatus.

through the membrane per unit area of membrane was plotted vs time. Results are shown in Figure 2.

From the data, both the release rates and diffusion coefficients could be calculated. By extrapolating the steady state condition to the x-axis, calculated from the following equation:

$$Dp = 1^2/6t_1 \tag{1}$$

The measured release rate (dQ/dt) was then calculated from the slope of the steady state line.

Since the measured dQ/dt did not take into consideration any effect due to boundary layer, and since measurable boundary layers were apparent in most cases, an equation derived at Rutgers was used for calculating intrinsic release rates:

$$dQ/dt_\infty = dQ/dt_{measured}/\gamma \tag{2}$$

where: $\gamma \quad = 1 - \dfrac{2(dQ/dt)}{Sh_R D_R C_s/d} m$

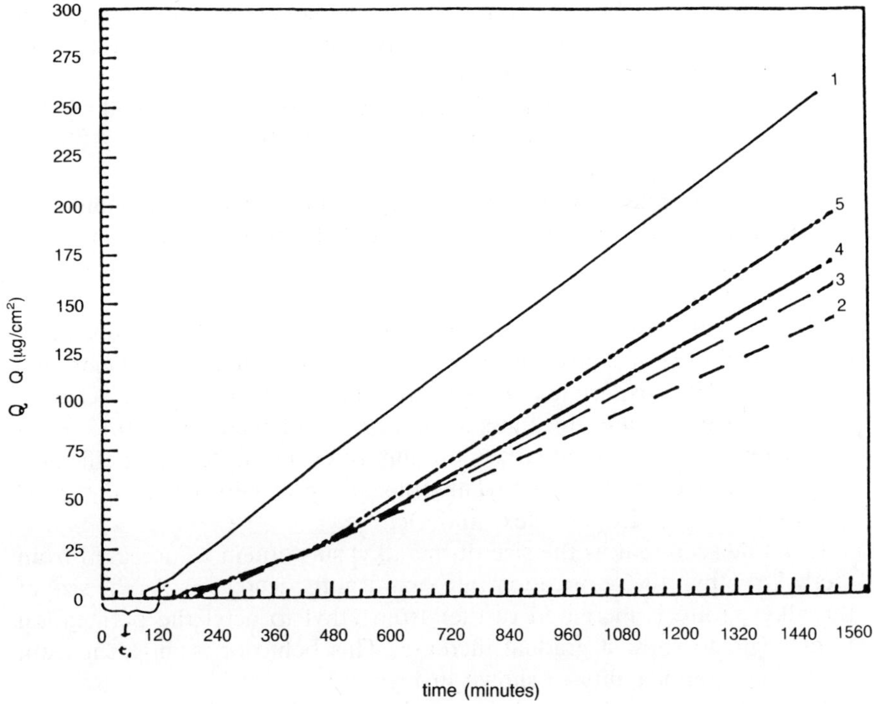

FIGURE 2 Progesterone permeation through $(MeRSiO)_x$ (1) $(Me_2SiO)_x$, (2) $(MeEtSiO)_x$, (3) $(MePrSiO)_x$, (4) $(MeHexSiO)_x$, (5) $(MeOctSiO)_x$.

and: dQ/dt_∞ = intrinsic permeation rate

dQ/dt_m = measured permeation rate

D_R = diffusion coefficient of progesterone in the receptor phase
= 9.81^*10^{-7} cm^2/s for progesterone in 40/60 v/v PEG 400/water

C_s = saturated concentration of drug in the donor phase
= 204 g/ml for progesterone at 37°C in 40/60 v/v PEG 400/water

d = length of the magnetic stir bar
= 2.54 cm for the G-C drug permeation cells

Sh_R = Sherwood number (calculated for the specific cell design, solute density, solute viscosity, stir rate, and stir bar length)
= 379 for the G-C cells with 40/60 v/v PEG 400/water at 425 rpms with 2.54 cm stir bar

The intrinsic release rates and diffusion coefficients thus obtained for the permeation of progesterone were found to be dependent upon the thickness of the membrane. This is illustrated in Table I for polydimethylsiloxane.

Under ideal conditions, the intrinsic release rates would be independent of the membrane thickness; however, the data shows that if the membrane is too thin, the experimental data are unreliable. To obtain accurate results, the membrane thickness should be maintained above 0.15 cm in this system. Table II summarizes the flux rates and diffusion coefficients of progesterone through methylalkylsiloxane polymers where the alkyl groups were: Me, Et, Pr, Hex, and Oct.

It can be seen that as the size of the alkyl substituent is increased from methyl to ethyl, the permeation rate significantly decreases. As the size of the alkyl group is increased further from ethyl to octyl the permeation rates begin to show a gradual increase. This behavior is different from that of gas permeability as shown in Figure 3.

In the latter case for example, as the size of the alkyl substituent is increased, the oxygen permeability markedly decreased, with $(MeEtSiO)_x$ being the most drastic.

This pattern is similar to that observed for the drug diffusion coefficient for methylalkylsiloxane polymers. The diffusion coefficient of pro-

TABLE I

Permeation Rates and Diffusion Coefficients for Progesterone Through Polydimethylsiloxane Membranes at 37°C and Various Membrane Thicknesses

Thickness (cm)	dQ/dt($\times 10^3$) (μg/cm^3-s)	γ	dQ/dt$_\infty$($\times 10^2$) (μg/cm^2-s)	dQ/dt$_\infty$1($\times 10^4$) (μg/cm-s)	diff. coef($\times 10^7$) (cm^2/s)
0.0305	10.40	0.30	3.43	10.50	1.03
0.0343	8.63	0.42	2.05	7.02	8.46
0.0381	9.62	0.36	2.71	10.30	9.43
0.1461	3.15	0.79	3.99	5.83	8.52
0.1473	3.03	0.80	3.80	5.60	8.87
0.2533	1.85	0.88	2.11	5.39	9.23
0.2578	1.97	0.87	2.27	5.85	9.12

TABLE II

Permeation Rates and Diffusion Coefficients for the Permeation of Progesterone through $(MeRSiO)_x$ Membranes at 37°C (R = Me, ET, Pr, Hex, and Oct)

Polymer	$dQ/dt_\infty I(\times 10^4)$ (μg/cm-s)	$Dp(\times 10^7)$ (cm^2/s)
$(Me_2SiO)_x$	5.67 ± 0.19	9.03 ± 0.29
$(MeEtSiO)_x$	3.34 ± 0.18	3.23 ± 0.24
$(MePrSiO)_x$	3.37 ± 0.03	3.35 ± 0.28
$(MeHexSiO)_x$	4.20 ± 0.30	3.09 ± 0.17
$(MeOctSiO)_x$	4.23 ± 0.31	2.43 ± 0.21

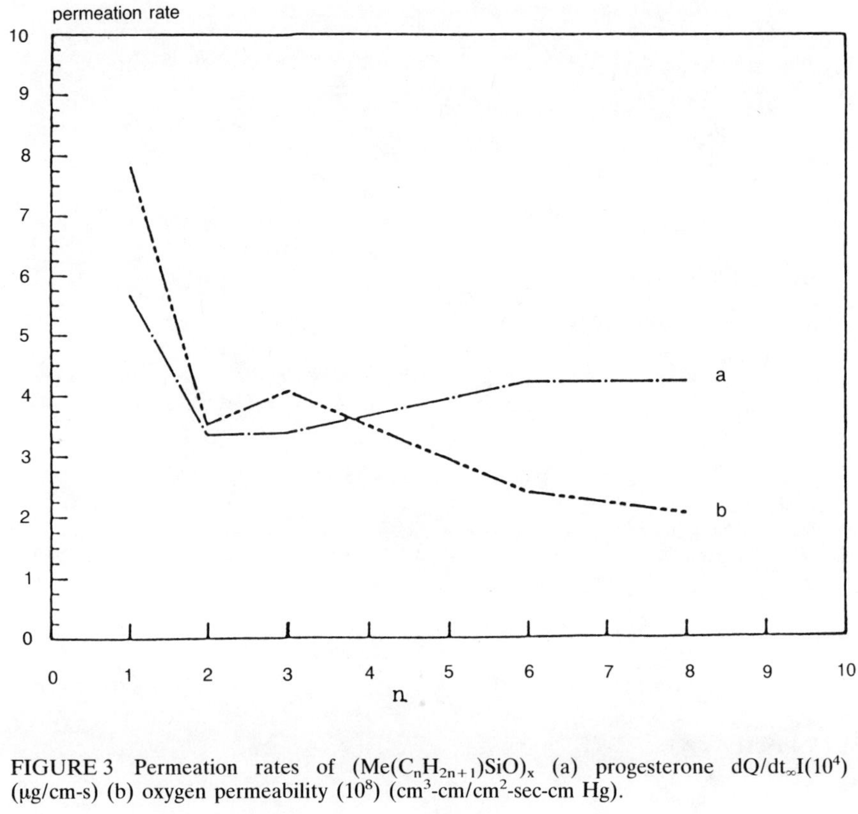

FIGURE 3 Permeation rates of $(Me(C_nH_{2n+1})SiO)_x$ (a) progesterone $dQ/dt_\infty I(10^4)$ (μg/cm-s) (b) oxygen permeability (10^8) (cm^3-cm/cm^2-sec-cm Hg).

gesterone through $(MeRSiO)_x$ decreases with an increase in the size of the alkyl substituent. The magnitude of decrease was greatest with $(MeEtSiO)_x$.

All of this data, along with additional data yet to be generated both at Dow Corning and Rutgers, will be used in designing a molecular model for predicting drug release rates through silicone polymers.

Index